THE ANCIENT MYTHOLOGY OF MODERN SCIENCE

THE ANCIENT MYTHOLOGY OF MODERN SCIENCE

A Mythologist Looks (Seriously) at Popular Science Writing

GREGORY SCHREMPP

McGill-Queen's University Press
Montreal & Kingston · London · Ithaca

ISBN 978-0-7735-3989-1

Legal deposit second quarter 2012
Bibliothèque nationale du Québec

Printed in Canada on acid-free paper that is 100% ancient forest free
(100% post-consumer recycled), processed chlorine free

McGill-Queen's University Press acknowledges the support of the
Canada Council for the Arts for our publishing program. We also
acknowledge the financial support of the Government of Canada
through the Canada Book Fund for our publishing activities. We are
grateful for additional support toward this book provided through
a Grant-in-Aid from the Office of the Vice Provost for Research of
Indiana University (Bloomington).

Library and Archives Canada Cataloguing in Publication

Schrempp, Gregory Allen, 1950–

   The ancient mythology of modern science : a mythologist looks
(seriously) at popular science writing / Gregory Schrempp.

Includes bibliographical references and index.
ISBN 978-0-7735-3989-1

   1. Religion and science. 2. Mythology. 3. Science news. I. Title.

BL240.3.S33 2012   201'.65   C2011-907843-0

Typeset by Jay Tee Graphics Ltd. in 10.5/13 Sabon

For Nina

Contents

List of Illustrations

Preface

A few weeks ago I took a walk around, looking for a metaphor with
which to end this book, a metaphor of a universe constructed, not by
a clockmaker standing outside of it but by its elements in a process of
evolution, of perhaps negotiation. All of a sudden I realized what I am
doing here for, in its endless diversity and variety, what I love about the
city is exactly the way it mirrors the image of the cosmos I have been
struggling to bring into focus. The city is the model; it has been all round
me, all the time.

Lee Smolin, The Life of the Cosmos (1997:299)

As a specialist in mythology I welcome Smolin into a long tradition of which he may or may not care to be a part, for the conviction that the human settlement models the cosmos – the root idea in
Smolin's metaphoric epiphany – is taken for granted in mythologies
the world over. Moreover, we have all encountered the present-day
mantra of "diversity"; and, more subtly, in much cultural analysis,
"negotiated" has emerged as a term of choice for summarizing the
basic condition of human society. One cannot imagine a more perfect moment to find out that the cosmos is a city brimming with
diversity and negotiation: the cosmos exhibits not only life (as promised nondescriptly in Smolin's title), but a form of life particularly
suited to our moment! Smolin's statement contains a declaration
of love and, along with this, a testimonial as to its source: Smolin
loves the city *because it mirrors the cosmos*. This literary cathexis
is not without self-regard, since the cosmos that the city corrobor-
ates is one that Smolin identifies uniquely with himself. In this case
and many others, what for the science writer is the discovery of a
powerful, absorbing cosmic image, is for the mythologist the discovery of yet another scientist discovering mythologizing. Smolin's
formula for popular science is a formula for mythology in general:

what the audience wants is an affective experience of correspondence between cosmos, settlement, and self.

Smolin presents the cosmos-as-city image as dawning upon him toward the *end* of his scientific treatise, an order of unfolding that is itself intriguing, for in the usual scenario it is myth that begets science rather than the other way around. Smolin's book is full of thought-provoking ideas. It also demonstrates what turns out to be one of the striking characteristics of popular science writing – specifically that this genre creates a novel context for images, devices, and themes that form the foundation of the human need to mythologize the world, a need that is far older than the efforts of modern science to describe the same world. The cosmos as city is no isolated instance, for across the board the images and strategies of persuasion that popular science writers keep rediscovering are those that have always propelled traditional mythologies.

☙

"It's just matter": this comment has stayed in my mind for over three decades now. I and a favourite uncle (a researcher at a pharmaceutical firm, and also the son of a Methodist minister) had been discussing the headline-grabbing story of an airliner that crashed in a snowstorm, leading the stranded survivors to eat the flesh of those who had died before rescue arrived. My uncle's serene initial response to the incident – "it's just matter" – reduced modern cannibalism to clinical truth. Almost immediately, however, he frowned and compromised this learned response with another, psychologically more primal, reaction. Noting that he himself probably would not have partaken of the meal, his reasoning was again terse: "it's taboo." In the space of several seconds, my uncle had tried out two opposing assessments – one obviously more professional, the other more personal (or at least intoned as such)[1] – encapsulated by the words "matter" and "taboo." While the man of science resolved a provocative quandary by categorizing the episode under a label – "matter" – that connotes moral distance and neutrality, the man of myth shuddered at the desperate breach of human decorum – a decorum that resists the levelling effect of this same label. The existence of taboos, their origin, and the consequences of their violation are popular themes in mythologies worldwide. Taboos are typically accompanied by a conviction that violations will produce cosmic repercussions, a belief which instances, in another form, the principle that the cosmos mirrors the settlement. By a modern society, "mythology" is typically regarded as the expression of a pre-scientific world in which the substance and structure of the cosmos are infused, irrationally, with deep human anxieties and hopes. "Science" connotes the opposite: in the interest of objectivity, the purging of such anthropocentric conceit from the idea of nature. The antinomy embodied in mythology and science has been felt profoundly by many individuals, and, continually evading closure, it has left a long trail in Western analytical thought.

Infused with this antinomy, popular science writing as a genre exhibits a strained hybridism of content and purpose. As writers like Smolin report the most exciting developments of contemporary scientific research, the ancient devices and strategies creep into their narratives even amidst proclamations of the triumph of science over myth. But while these devices are ancient, the context into which they are deployed is new and allegedly modern. We encounter, in popular science writing, not a mythologizing of the cosmos as it was perceived by purportedly naive, archaic peoples (an idea that has given rise to countless fables about our intellectual origins), but rather – and this is what makes the venture interesting – an attempt to mythologize the cosmos of matter, that is, the cosmos of de-mythologized, impersonal substance.

⚮

Popular science writing responds to a broad market for intelligent but simplified accounts of complex theories in realms of inquiry that many readers left behind with their formal schooling. In my case, an early interest in the sciences, and especially mathematics, gave way to a graduate-level specialization in mythology. From this vantage, popular science writing takes on special relevance, for in experiencing the cornucopia of views about the origin and nature of the cosmos displayed in traditional mythologies, the mythologist cannot help but wonder what, by comparison, the modern sciences have to say on such topics. At some point my specialized interest in mythology interacted with my more general interest in the modern sciences, the latter fed increasingly by popular science writers. The interaction of these two interests gave rise to musings about popular science writing itself as a form of mythology. This book elaborates on those musings.

ACKNOWLEDGMENTS

I feel gratitude of mythic proportions to four colleagues who provided help that was crucial and generous beyond all expectation in various phases of this project: William Hansen, Joseph Nagy, Marshall Sahlins, and, throughout, my wife Cornelia Fales.

A number of other colleagues have provided critical help at many points and in many ways: Richard Bauman, Linda Dégh, Raymond DeMallie, Sandra Dolby, Gary Downey, Hasan El-Shamy, Christina Fales, Henry Glassie, Lynn Hooker, Wally Hooper, Judith Huntsman, John Johnson, Liz Locke, Jens Lund, Emily Lyle, Moira Marsh, Portia Maultsby, John McDowell, Daniel Peretti, David Shorter, George Stocking, Ruth Stone, and Sue Tuohy.

I have tried out my arguments in a variety of professional presentations, informal gatherings, and classes I offer on mythology and comparative cosmology at Indiana University; and I have benefited from responses too numerous to attempt to list.

Finally, I am grateful to Kyla Madden and the staff of McGill-Queen's University Press, as well as to the anonymous readers, for their many efforts and help in getting the final manuscript in shape.

THE ANCIENT MYTHOLOGY OF MODERN SCIENCE

Centaurs and Scientists

Two observations converge in the idea of this study. First, books in which accredited specialists discuss science and/or specific scientific findings in terms accessible and appealing to the general public have come to occupy a prominent place in contemporary bookstores. Second, humans throughout history have been fascinated by mythology. My argument is that these two circumstances are not unrelated; or, more specifically, the thesis of this book is that popular science writing provides a primary arena for the creation of contemporary mythology.

Popular science writers typically use the term "myth" (or "mythology") to designate just those pre- or non-scientific understandings of the world from which they will attempt to wean their readers.[1] In invoking myth in order to dismiss it as an inferior mode of understanding, these contemporary writers carry forward a long intellectual tradition that began with classical Greek philosophers and was given renewed emphasis by Enlightenment thinkers – whence it passed to nineteenth- and twentieth-century social thought. Despite shifts in nuance, the alleged defect of myth, from the Greek philosophers to the present, has remained constant: myth portrays the cosmos anthropocentrically, that is, *not as it truly is, but in our own image – skewed by human passion, parochialism, and self-infatuation.*

Fragments from the presocratic philosopher Xenophanes are often held up as the first articulation of a critical attitude toward anthropocentric bias: "The Ethiopians say that their gods are snubnosed and black, the Thracians that theirs have light blue eyes and red hair. But if cattle and horses or lions had hands, or were able to draw with their hands and do the works that men can do, horses would draw the forms of the gods like horses, and cattle like cattle, and they would make their bodies such as they each had themselves" (from Kirk et al. 1983:169).[2] For reasons that we will explore, Copernicus in the modern world comes to displace Xenophanes as the hero who pointed out and problematized this human inclination, but the verdict on myth remains unchanged: it forms the exemplary case of anthropocentric illusion.

With this historically resonant verdict in mind, popular science writers typically claim to offer their readers an understanding of the cosmos that transcends the fatal flaw of myth. I am skeptical of this claim.[3] I will argue that popular science writers, in attempting to make the findings of science humanly compelling, fall back on or reinvent in sly and subtle variation the anthropocentrically motivated devices of traditional mythologies. Because so many popular science writers are steeped in the myth/science dichotomy and its rhetoric, my claim that popular science writers themselves are producing mythology will no doubt be taken as a challenge, if not an affront, by many devotees of this genre.

Yet I also foresee the possibility of a very different reaction to my thesis, one that greets it as obviously true. Some of the previous scholarly critics of this genre have tapped the term "myth" or "mythic" impressionistically and typically rather loosely;[4] for these, my thesis may be unsurprising. A different form of acquiescence, one that I would not especially relish, might come from those who invoke the term "mythology" more radically to refer to any set of fundamental assumptions about the world; science from this point of view is one mythology among others. This usage is as problematic as the renunciation of myth proclaimed by popular science writers in characterizing their own product; in place of uncritical dichotomization of myth and science, we are offered uncritical conflation.

The epistemological relativism lurking in the latter notion of science – that it is just one more mythology – figured importantly in the recent so-called "science wars";[5] and since these wars have so greatly influenced the tone of contemporary debate about science and culture, I will say a word about my own position. At one extreme in these debates lies the view of science as a type of cultural construct – accompanied, sometimes, by the claim that this construct is rooted in particular class interests and/or globally hegemonic ambitions. At the other extreme lies a vision of science as a progressive, culture-transcending venture that promises final, pan-human validity. For purposes of this book, I claim neutrality on such issues, for two reasons. The less important is that I am ready to discount neither view entirely. The more important is that the main arguments of my book are not directed toward science at its core; rather, in focusing on the popularizing of science I attempt to pick up science at those points, admittedly beyond precise location, at which it starts to become – dominantly – something else, a something else that by virtue of my background in mythology I think I recognize.

What I offer in this book, both to those at the outset inclined to agree and to disagree with my thesis, is a counter to the polemical and/or artfully vague invocations of the idea of myth that have dominated debates about science and culture. In such invocations, whether in dichotomizing or conflating mode (as discussed above), the term "myth" has served less to initiate critical analysis than to cast a spell. As a specialist in comparative mythology,[6] I cannot help but be intrigued, and in a sense delighted, by the popularity and magical power of this term; but I also look for precision and concrete example. I do not see a monolithic definition of "myth" as a useful or even attainable goal, for "myth" is, and will likely remain, a "fuzzy-set" of a term. Still, a fuzzy-set implies specific, recurrent traits or qualities that, even though combined flexibly and inconsistently in practice, can individually be subjected to careful analysis – to which I turn in my first chapter. The chapters that follow then focus in detail on unravelling specific mythologizing strategies in the arguments of a variety of popular science writers. A force that popular science writers decry even while trying to exploit, myth offers unique entree to their craft and, more broadly, to a number of the overarching concerns of contemporary thought: the nature of human cognition and communication, the place of science in contemporary culture, and the human search for sustaining moral visions.

൭

This book has six chapters. In chapter 1, I elaborate on the basic concepts used in this book, especially myth and popular science writing. I emphasize the history of the idea of myth in relation to centric (anthropo-, ethno-, phallo-, etc.) illusion, focusing on highlights of direct relevance to popular science writing. The remaining chapters (2 through 6) explore the wide range of strategies through which science can be mythologized, each chapter drawing together some of these strategies, along with analysis of various writers' use of them, under one or a few organizing example(s).

Perhaps the most basic way of injecting mythological power into science is to recount, in the guise of science, a story that the audience already knows as a myth. Chapter 2, my first case study, considers how astronomer John Barrow recasts the story of Prometheus (without ever mentioning this hero) through quantitative analysis and a discourse on the constants of nature. I explore the ways in which Barrow uses a fire-domestication scenario, subliminally tapping its mythic appeal, in an attempt to bolster his position on the nature-nurture debate – a modern inflection of the moral concern with freedom vs. constraint that has always accompanied the Prometheus myth.

Chapter 3 considers the strategy of persuasion through microcosm/macrocosm analogies, especially as utilized by the late Stephen Jay Gould in *Full House*, the book he called his "most beloved child." Given that Gould's big point is to challenge the idea of progress in biological evolution, the reader may be puzzled to find that much of his book is taken up with seemingly unrelated issues, including an inspiring story by Gould about his own early battle with cancer, and an analysis of the decline of baseball batting averages. Prompted by the knowledge that in traditional mythologies the life of the body and culturally focal athletic contests typically are regarded as having cosmic correlates and consequences, I examine the many ways in which Gould uses these and other favourite microcosms to seduce readers into his larger views regarding biological evolution and our place within it. Also among his microcosms is "the drunkard's walk," a scenario often used as a cautionary tale about statistics. Juxtaposing Gould's vision and that offered by Leonard Mlodinow in his recent *The Drunkard's Walk*, I offer some concluding reflections on randomness and contingency and how these themes can be played to popular effect.

If chapters 2 and 3 deal with the grand sweep of evolution, cosmic and biological, chapters 4 and 5 deal with topics closer to home. Specifically, they deal with issues of empirically based cognitive research, one of the more robust arenas of contemporary popular science writing. In traditional mythologies, anthropomorphic images are projected directly onto the cosmos; most notably, the cosmos is typically conceived of as a gigantic kin-group, organized into sub-families (stars, humans, animals, plants, rocks, bodies of water, and so on, often with Sky and Earth as parents). Starting with the use of kinship metaphors by modern theorists reflecting about the genesis and structure of "the category," I argue in chapter 4 that the category offers an alternate route for cosmically ambitious anthropomorphic desire: conceived of as mechanism of all knowledge, whatever one puts into the concept of the category, one thereby puts onto the cosmos. My central examples are from category theorists George Lakoff and Mark Johnson, with special emphasis on their *Philosophy in the Flesh*; but I will also discuss Lakoff's *Whose Freedom?* and other works in which he carries his interest in metaphor into political punditry, particularly in the run-up to the most recent American presidential elections. Some readers will be surprised to find Lakoff and Johnson treated in a book about popular science writing; however, their work reflects the main characteristics of this genre in its present state.[7] To pass over their work because it focuses on metaphor rather than planets or amoebae betrays a parochialism characteristic of this genre's past, but not of its current trajectory (that trajectory is considered in chapter 1); indeed, application of various cognitive science perspectives to moral/political reasoning is a major growth area of contemporary popular science writing.

Chapter 5 looks at what happens when our anthropomorphic inclinations turn inward in another, but very different, sense – that is, when we portray processes such as reasoning and/or performing specific intelligent tasks as routines directed and/or carried out by a society of little inner agents, or "homunculi." Fully aware that appeals to the "little man inside" elicit the charges of "folk psychology" and "homunculist fallacy," some artificial intelligence researchers nevertheless see in homunculism a useful heuristic and research strategy. Focusing on the work of Daniel Dennett, I juxtapose the introversive anthropomorphism of the homunculus with the extroversive anthropomorphism typically labelled as mythology, in order to draw out a more precise understanding of the basic mechanisms of anthropomorphizing and our inclination to it. Homunculism and its critique paradigmatically illustrate the logic and pitfalls of popular treatises on a variety of internal physiological structures – including genes, memes, cells, organs, ova, and sperm – for which we are tempted to over-attribute characteristics of the human whole to a particular part of that whole.

Beginning with Steven Weinberg's classic *The First Three Minutes*, chapter 6 explores different uses that popular science writers have found for the Copernican Revolution, and the creative variations that have arisen in portrayals of it. The main focus in this chapter is an intriguing side-episode from the Apollo moon missions, one propelled by the idea that the view of Earth rising from the lunar horizon (the vision of "earthrise") offered a new – less mathematical, more visceral – illustration of Copernican spatial relativity, a lesson about geocentric/anthropocentric illusion that would be graspable by the masses. I discuss several writers' renditions of this idea, and, with Carl Sagan as central example, show how Copernican decentring is invariably accompanied, if not sabotaged, by attempts to reconstruct a sense of the cosmic exceptionalism of our species. The continuing, if not growing, influence of this exceptionalist vision is evident from many sources, including the moon and space exploration nostalgia books that currently flood the popular market.

I conclude by drawing together these chapters in a more general commentary framed as paradise lost and regained – a scenario that occurs in the mythologies of many societies and that furnishes a recurrent underlying narrative for much popular science writing. The consequences of science are mixed: greater maturity and control of nature are accompanied by forebodings of the loss of emotional consolations that mythology and religion have supplied. I explore the ways in which popular science writers attempt to offer compensation for these losses, and redemption through resolution of ambivalence. I also consider tensions and ironies in their project, notably the possibility that attempts to make science culturally and politically relevant will entangle it in particular ideologies or worldviews – a state of affairs these writers are quick to decry, repeatedly citing this hazard as the lesson of the Copernican Revolution.

☙

The case study chapters (2 through 6) of my book are organized to conform loosely to the heroic story of science they analyze; recall, for example, the indelible version of that story offered by the late Stanley Kubrick (*2001: A Space Odyssey*) through a sequence that moves swiftly from the first use of a tool to space explorers poised on the outpost of the moon. The subject of my chapter 2 is the acquisition of one of those first tools – fire – and the claim that it opened for our species the possibility of accelerated mental and cultural evolution. Chapter 3 also considers the evolution of life on Earth and popular beliefs about human ascendancy within this process. Chapter 6, my final chapter, sees humans standing on the moon and looking back at Earth, even while poised to colonize the cosmos. The chapters between (4 and 5) deal with the growth of knowledge within the brief gap between these two moments, and again reflect a standard heroic order: nature mythology is displaced by the abstractive tools of philosophy, and science culminates in the explanation of the workings of the organ of its creation, the human mind.

One will notice that all of my chapters focus on entities that are small: a Tom Thumb fire-maker, various *micro*cosms, the "category" (the "atom" of the cognitive world), homunculi, and finally a recent but small Copernican Revolution – one inspired not by the greater light that rules the day but the smaller one that rules the night. This slant toward the small is one indication of the subtlety of the mythologizing that goes on within popular science writing, where it *must be* subtle, since in the context of science, myth is taboo. This pattern stands in contrast to archaic mythologies, which nearly always portray our progenitors as larger and more long-lived than we are today. While it is not without reason that we invoke the phrase "mythic proportions" to mark "something big," the analyses that follow will suggest that proportions need not be large to be mythic.[8]

Besides the *large*, this study also veers away from mythologizing that is *obvious* – for instance, in debates surrounding the so-called Gaia hypothesis, or the proposal that Earth, with its ecological feedback mechanisms, is a self-regulating, living organism.[9] These debates are so saturated with traditional mythological allusions that biologist Massimo Pigliucci pleads: "So, please let Greek gods and goddesses rest in peace on Mount Olympus, and let us get back to developing a much-needed real science of planetary ecology" (2005:26). Fascinating in its own way, such "upfront" mythologizing is not the main topic of my book.[10] I want to show that mythical arguments and strategies are to be encountered even in expositions of science that invoke little or nothing of the tangible content of traditional mythologies: no Gaia, no Zeus, no centaurs!

But given that the kind of mythology I consider has no real need for centaurs, I owe the reader an explanation for my summoning of this creature in the introduction's title. Just as Smolin and his mythically minded predecessors find the character of the cosmos in the image of the city (see preface), so the centaur captures the fundamental quality of popular science writing that I address in this book.

As much as any one image can, the centaur calls to mind what one might term the spirit of mythology, a force that subtly but insistently infuses the science that is offered in popular science writing. The appeal of the centaur surely has something to do with the troubling hybridism of this image, the crossover between qualities regarded as contrary if not mutually repelling. Popular science writing is a hybrid craft,[11] one that combines different conventions of presentation as well as divergent ways of thinking and feeling about the human place in the cosmos, often conjuring the hope that disparities might be resolved or transcended. Moreover, among the energizing dichotomies engaged most frequently by contemporary popular science writers is the ancient one posed by the centaur: between the qualities we like to claim as our unique character, our humanity, on one hand, and the characteristics that we share with other entities of the cosmos, our animality and materiality, on the other.

A recent, striking example of the engagement of such dichotomy is found in Lynn Margulis' popular work *Microcosmos: Four Billion Years of Microbial Evolution*. Margulis challenges traditional ways of thinking about human uniqueness by arguing for the microbial origins of the cells that combine in higher organisms. Her reflections about biological combinatory processes also bring Margulis to the point of hinting at combinatory play between science and mythology:

> Human religion and mythology have always been full of fantastic combinations of creatures – the mermaids, sphinxes, centaurs, devils, vampires, werewolves, and seraphs that combine animal parts to make imaginary beings. Truth being stranger than fiction, biology has refined the intuitively pleasing idea with its discovery of the overwhelming statistical probability of the reality of combined beings. We and all beings made of nucleated cells are probably composites, mergers of once different creatures. The human brain cells that conceived these creatures are themselves chimeras – no less fantastic mergers of several formerly independent kinds of prokaryotes that together coevolved. (1986:120)

Margulis here opposes mythology and science even while aligning them in the claim that science refines mythology's "intuitively pleasing idea." To assert that the amazement and pleasure readers seek in mythology will be met even more amply by science – that science offers "the greatest show on earth" (in Richard Dawkins' phrase, presumably borrowed from the circus)[12] and will beat mythology at its own game, so to speak – is a common strategy of popular science writers, one that we will encounter in novel variations as we proceed. At points, the interaction of science and myth itself might almost be described as a form of coevolution.

Two millennia ago, the Roman poet Lucretius, a sort of "patron saint" for popular science writing,[13] also ruminated on centaurs, seeking to account for them through a physicalist, atomistic rendition of optical illusion. He reasoned that microscopic films cast off by humans and horses collide and stick together before striking the eye, creating a false sensation (4.720ff. [1951;152ff.]). Two millennia later, David Hume, in his great *Enquiry Concerning Human Understanding*, also invoked the image of the centaur to account for the mind's capacity to create "fictions" (2007:34); and nineteenth- and early twentieth-century mythologists, led by E.B. Tylor and James Frazer, drew from Humean epistemology – except for the overall skepticism, which they conveniently ignored – the dichotomy of science vs. myth that nowadays inhabits popular science rhetoric. Like the centaur, contemporary popular science writing arises out of a swirl of ideas and images cast off from different sources and reflecting different human capacities and concerns. These fuse in creative and appealing, though often ultimately illusory, combinations. A mythologist's tools are well suited for an analysis of these compounds, one going beyond polemical dichotomies and offering a unique window on the interaction of science and myth in the cultural sphere. Such an analysis is the ambition of this book.

Mythologizing Matter: On Myth, Popular Science, and the Problem of Centric Knowledge

The purpose of this chapter is to elaborate on the key concepts already introduced – popular science writing, myth, and anthropocentrism – in order to lay groundwork for the chapters that follow. I also discuss some highlights from the history of the idea of myth, focusing on developments of particular relevance to contemporary popular science writing.

POPULAR SCIENCE WRITING

By "popular science writing" I mean non-fiction works written by accredited specialists with the aim of presenting scientific theories and findings, and/or general views about the nature and present state of science, in an accessible and appealing way. Typically the authors of such works adopt broad, integrative intellectual and cultural goals, which might include synthesis of scientific and literary values, interdisciplinary dialogue, the connecting of science to moral and political issues, and/or the construction of compelling narratives about science and its place in history.

A number of difficulties surround the characterization just offered, and elaboration is in order. Among the writers I will treat in this book, centrally or in passing, are John Barrow, Stephen Jay Gould, Steven Weinberg, Carl Sagan, Daniel Dennett, Stephen Hawking, Richard Dawkins, E.O. Wilson, and Steven Pinker. The most widely used genre terms for the works of such writers – or more precisely, those of their works oriented toward a general audience – are "popular science" and "science writing." Both terms are problematic. To some, popular science carries an "amaze your friends" connotation that is largely inconsistent with the works considered here, while science writing is sometimes used to designate either specifically journalistic writing or technical writing for specialists. Further, practising scientists sometimes resist adopting either popular science or science writing as labels for their own works, preferring to reserve them for authors whom they consider to be writers first and scientists second. These terms are far from ideal and one cannot fail to notice discontent with them among practitioners; but none of the several proposed alternatives has as yet gained wide acceptance.

Yet for all of their deficiencies, these terms do have one big attribute to recommend them: people know them. I have chosen the amalgamated term "popular science writing" in order to broadly alert potential readers about the subject matter of this book. Once beyond this introduction, I will often, without change of meaning, shorten the phrase to "science writing." Most of the works I mention in this book, and all that I focus upon in detail, are written by practising scientists, and they lie at the serious and seriously accredited end of the continuum of works that might be catalogued under popular science or science writing.

◌◦◌

An alternative to these terms is currently being proffered by John Brockman, the brilliantly successful literary agent for many, if not most, of the best-known popular science writers. Brockman proposes as umbrella term "the third culture." This term stems from the British critic and academic novelist C.P. Snow, who in 1959 coined it to designate what he saw as a great rift between scientific and literary culture. Snow suggested the possibility of a "third culture," which would combine the two, and pointed toward social history as offering this possibility (1993:68ff.). Brockman's third culture intellectuals come from the sciences; and, departing from Snow, Brockman portrays the synthesis as a takeover: "Throughout history, intellectual life has been marked by the fact that only a small number of people have done the serious thinking for everybody else. What we are witnessing is a passing of the torch from one group of thinkers, the traditional literary intellectuals, to a new group, the intellectuals of the emerging third culture" (1996:19).[1] Descriptively empty and at odds with its original meaning, the term "third culture" as used by Brockman is not useful as a genre label, whatever its promotional value.[2]

But if his proposed label does not help us, Brockman's vision of this genre does. Brockman notes that such writers are not *merely* popularizing, i.e., writing downward to the masses. They are also attempting to share insights and perspectives with neighbouring specialties, portending a wide interdisciplinary discourse.[3] The rise of works that attempt to speak simultaneously to the general public and interdisciplinarily inclined academics also contributes to the strain on the designation "popular science writing," which, as noted earlier, I adopt provisionally while acknowledging that my usage sometimes stretches traditional associations.

But interdisciplinary discourse is not the only goal; even the brief passage by Brockman quoted above alludes to at least three other ambitions that are embraced by most contemporary popular science writers.[4] All three are of immediate interest from the standpoint of mythology. First, one repeatedly encounters the conjoining of the two cultures, science and art, as a goal of popular science writers; indeed issues of imagination and aesthetics that were once peripheral are steadily moving toward the centre of this genre. In his well-known work *Consilience*, biologist E.O. Wilson, for example, says, "The greatest enterprise of the mind has always been and always will be the attempted linkage of the sciences and humanities" (1998:8).[5] Virtually every successful work of contemporary popular science writing strives to incorporate an artistic dimension. The attempts range from the explicit thematizing of the goal, as in Wilson, to modest efforts to write gracefully and to inject a note of "humanity" here and there. We will consider a number of such attempts in the chapters that follow.

Second, Brockman promotes the idea of scientists as "intellectuals," implying their readiness to engage issues of public cultural, moral, and political significance. These new intellectuals will thus transcend not only the divide between science and art but as well the related divide between "fact" and "value"; one potent use of science is precisely to create a high ground from which to offer insights and recommendations on moral and political issues. These insights range from ruminations about the perennial big questions – the meaning of existence or the origins of morality, for example – to contemporary social issues, such as the long-running nature vs. nurture debate. In the chapters that follow, we will similarly encounter these and many other moral issues woven into the exposition of science.

Third, Brockman's scenario of "passing of the torch" amounts to an episode of what is sometimes called the heroic story of science. As much as any other human enterprise, science needs stories – to portray its origin and development, to glorify its heroes, to dramatize its methods and values, and to proclaim its place in human civilization. Virtually all popular science writers make use of storytelling; their stories range from humble first-person anecdotes, to hero-tales, to grand epics of human and cosmic evolution.[6] Traditional mythologies offer numerous templates for the construction of narratives of triumph; and usurpations, especially of an older line of gods by a younger one, are among the most favoured of such scenarios. The passage by Brockman quoted above gives us just such a scenario: the older literary Titans falling before the new scientific Olympians of public intellectual life.

The potential for mythologizing opened up by these three integrative ambitions should not go unnoticed. Art is often inspired by and/or imbued with mythology. Cultural, political, and moral values are at the core of mythological stories and symbols. Narrative is an indispensable and basic – perhaps *the* most basic – vehicle of persuasion. To the extent that popular science writers attempt to infuse science with art, moral values, and/or narrative persuasion, such writers open the door for science to metamorphose into myth.

QUALITIES OF MYTH

The terms "myth" and "mythology" often designate stories of supernatural beings or heroes set in ancient times and telling of cosmos- and society-shaping deeds and events: the origin of the seasons, the reasons why we must die or that we are incapable of understanding the speech of animals, or the founding of basic social mores and taboos. A number of connotations of myth stem from patterns or structures of coherence that typically arise in such mythological stories or that form the backdrop – the worldview – of such stories. Mythical portrayals of the cosmos, for example, revel in microcosm/macrocosm parallels. Both large- and small-scale structures are typically portrayed in mythology, often no doubt unconsciously, as recapitulating the village; the design of the cosmos reflects the local human settlement, but so may the social organization of fish in the depths of the sea. The human settlement is typically portrayed, moreover, as lying at cosmically privileged coordinates of time and space. The various archaic mythologies of the world arise from sophisticated observations about the empirical world, shaped through imagination, moral speculation, and anthropocentric desire.

In other usages of "myth," however, ancientness is less important: myth includes even modern stories, images, or ideals when these rivet minds, energize cultures, and confer identity, design, and/or destiny on individuals or societies. One might hear, for example, of the "myth of the frontier" or the "myth of the open road" in the American mindset. Such usage replaces temporal primordiality with a kind of psychological primordiality, or ultimacy, and may yet carry hints of the numinous and the timeless.[7] All of the above usages to some degree impute to myth a certain kind of social and psychological function: that of providing societies and individuals with an ultimate moral ground, a validation and justification of an accepted or prevailing way of life and scheme of values.

Recurrently, one encounters divergent views in theories of myth about the relation of myth to narrative. Some scholars insist myths must be stories; others are not so restrictive. As implied above, certain ways of structuring the cosmos (e.g., as a large village, or as an encompassing kin-group in which the elements of the cosmos descend from cosmic parents such as Sky and Earth) are often said to be mythic in fundamental character. Where such structural conceptions exist, they will often have associated narratives; but who is to say which is primary: does the story give rise to the structure, or does the structure generate the story? Whether myths are necessarily narratives has been a divisive issue in theory of myth. I should therefore be clear that I am sympathetic toward many non-narrative approaches to myth and, in some contexts, will be talking about myth in which no overt story is present.

Millennia of philosophical and political debate about myth have given rise to contradictory insinuations regarding its veracity, and the term has taken on complex and contradictory valences of deeply true and profoundly false.[8] Connoting in some instances a synthesis of poetic imagination and moral wisdom held as a society's unique spiritual treasure, "myth" or "mythology" just as readily designates ideologies perpetrated by demagogues upon gullible masses. Vexingly, such opposed possibilities are not always distinct or mutually exclusive. Popular usage often sides with the false pole of the dichotomy: virtually any widely and/or deeply held misconception is a candidate for the label "myth." For some, the term "myth" points to a tension, perhaps a deep flaw, in our cognitive apparatus, specifically an inclination to privilege narrative syntheses that are emotionally moving over those that are rigorously responsible to known facts.[9] Finally, myth or mythology can also signify organization or coherence more generally, as a *set* or *totality* of elements integrated around a compelling theme and comprising something like a worldview. One hears of Greek mythology, but nowadays also of "Elvis mythology" or "Star Trek mythology."

Since a few, or even one, of the many qualities just noted can precipitate use of the term, we are left with a trail of varied but overlapping qualities, all connoted by "myth." Each of these is tapped into by one or more of the popular science writers that I consider in this book; I bring each quality into focus as it becomes relevant, along with theoretical perspectives addressing that quality from the tradition of myth study. No doubt some readers will be perplexed if not frustrated by my fuzzy-set approach, that is, by my setting out a group of diverse qualities that in varying constellations (or sometimes even singly) elicit the term "myth." However, my experience leads me to insist that, given the many partially overlapping usages, most of which have some merit, no pat definition of myth will suffice; certainly none of the pat definitions of myth proposed thus far by scholars has risen to general acceptance.

Yet there is a way in which the qualities just discussed can be unified analytically. Specifically, all of these qualities can be explored together under the umbrella term of anthropocentrism, for collectively they are the modalities through which anthropocentric visions take shape. Myths usually do proceed as if we humans are at the centre of the cosmos (figuratively if not literally) and as if the cosmos is specifically "for us" and slanted to our interests and needs. That we are the centre (literally or figuratively) and that the cosmos is "for us" are the fundamental assumptions of an anthropocentric purview.

It is sometimes useful to distinguish another concept related to anthropocentrism, namely anthropomorphism – the assumption that the other elements of the cosmos are "like us" in form, character, or organization.[10] "Anthropomorphism" can be defined either narrowly or broadly. Narrowly, it means the imposition of human form on a non-human object – for example, a clock whose "face" has been overlaid with features of a human face. But we also frequently use "anthropomorphism" more broadly to designate the projection, onto non-human objects, of designs or artifacts that humans have created as extensions of themselves. To proclaim that the human settlement mirrors the cosmos is also to anthropomorphize the cosmos, by further extending a human extension, so to speak.

Usage of "anthropomorphism" is often broad in another sense as well. It is one of the quirks of this term that although the "morph" root connotes form, we usually use "anthropomorphism" to designate transference of just the opposite, i.e., *intangible* human properties such as intelligence, personality, or mood – as occurs, for example, when an individual talks to an automobile. As will be evident through this book, my use of "anthropomorphism" is broad in both of these senses. Anthropocentrism and anthropomorphism are almost always mutually implicating in myth. For purposes of this book I follow the typical usage of considering anthropocentrism to be the broader term, often invoking it as the default to cover both. I approach anthropomorphism as a variety of anthropocentrism: the assumption that other entities of the cosmos have us as model flows from the conviction of our own centrality, and it implies that we can relate to and manipulate them on our own terms.

As noted above, the many specific qualities connoted by "myth" can be gathered under the concept of anthropocentrism (which, as just noted, in my usage includes anthropomorphism). The most obvious example is the above-mentioned mythic predilection for microcosm/macrocosm parallels, such as that between cosmos and settlement: clearly this predilection stems from the broader inclination to anthropomorphize the cosmos, and thus make it familiar. Or consider the narrative quality of myth – the *story*. It is often argued that narrative is a modality of understanding specific to human consciousness, a sense of meaningful unfolding imposed by humans on, rather than an intrinsic property of, cosmic process (Stephen Jay Gould's version of this claim will be considered in chapter 3). Hints of a plot in cosmic or human evolution ipso facto raise suspicions of anthropocentric intrusion. Although, as noted above, I do not limit myth to narrative, popular science writers frequently do engage in storytelling, and the issue of anthropocentrism must be considered in such instances. Popular science writers also revel in the creation of metaphors, and metaphors can be anthropocentric in many different ways. Narrative and metaphor, and the theoretical issues connected with these, will arise repeatedly and in varying forms in the chapters that follow. But suspicions of anthropocentric intrusion must similarly arise for virtually all of the other specific qualities of myth cited above: if not anthropocentric in principle, they are easily co-opted by whatever anthropocentric inclinations we bring to them.

In the context of science, "myth" typically takes on an overarching connotation of anthropocentric bias. Convinced of the sophistication of mythical thought, unconvinced that mythical thought can be entirely transcended, I nevertheless concede the general accuracy of this connotation, and adopt it as a theme in the chapters that follow. Mythologizing by popular science writers sometimes is a rather superficial affair, a matter of mere packaging. In the instances I consider in the following chapters, the mythologizing cannot be dismissed so easily.

MYTH AND POPULAR SCIENCE WRITING

Since the Enlightenment, myth has been thought of as a first stab at explaining the workings of the cosmos, a "proto-science." Though disinclined to admit to it, popular science writers often invert this sequence, giving us science that aspires to be myth. But on closer inspection, these two formulas – myth striving to be science, and science striving to be myth – turn out to be not quite reciprocal. For contemporary popular science writers do not seek to reverse – generally they staunchly uphold – the Enlightenment view of myth as an early, naive stage of human knowledge, and of science as the passage to a higher stage. In the end popular science writing constitutes not so much an inversion of the Enlightenment formula, as the addition of a third state that has resonances with the first: myth gives way to the findings of science, and then popular science writers use those findings as a base from which to remythologize the cosmos.[11] In doing so they reincarnate, in a modern and popular form, a dialectic that in rudiments is by now quite old, one perhaps set in motion the moment someone first envisioned an understanding of the cosmos superior to that offered by myth.[12]

One should not underestimate the number of factors that have contributed to the rise of popular science writing in our time. These include the increasing influence of science and technology in contemporary life, and a demand for accountability for them; the politics of science-funding; the public appetite for new metaphors, icons, and objects of fascination; the broadening of celebrity book culture; and the growing demand for more immediate relevance and an increased entertainment quotient in education generally. A scientist may also turn to the popular arena, or the court of public opinion, out of frustration with some aspect of the professional theory, practice, or organization of science. But such diverse factors converge in a fundamental goal toward which many popular science writers, in complicity with their audience, strive: an authoritative vision of the cosmos offering insight into "the big questions," those first posed in archaic mythologies. What are we? What is our origin and place in the cosmos? What is the good? Does the structure of the cosmos hold a message for us or about us? On such matters, the individual devotee of this genre can accept wholesale one particular writer's scientific vision, or attempt to piece together a collage, either in parallel with or disaffection from the claims of other works and traditions that proffer cosmic delight and wisdom (mythico-religious, philosophical, artistic, New Age, even self-help). But the appeal is to the collective as much as to the individual: popular science writing offers a venue through which a writer can disseminate a message broadly and develop a large following. Such demographic ambition itself forms one more point of convergence between popular science writing and myth, for the latter too carries a connotation of broad dissemination.

In an interesting twist, however, the connotation of broad dissemination is most visible in the everyday use (often adopted by popular science writers) of "myth" to mean "falsehood," for a speaker is likely to invoke the term "myth" over "falsehood" (or "error" or "lie") precisely when he/she regards the falsehood in question to be widespread and intractable. By virtue of this connotation, myth emerges as rhetorical foil for those intent on disseminating truth widely, even while the demographic success of mythological stories cannot but inspire imitation. It is important to note, however, that "myth" or "mythology" as falsehood, a usage favoured by popular science writers, can be misleading. It is not clear that at the times and places of their creation even the oldest mythologies had any intention other than to honour the best-available empirical observations of the natural world. The distinguishing principle of myth (vis-à-vis science) should not be sought in an indifference to empirical reality on the part of myth. It should be sought rather in myth's readiness to imaginatively *leap beyond* limited empirical observations and set sight on grand humanistic and aesthetic visions.

Analogously, the distinguishing principle of popular science writing should not be sought in mere simplification, for this gives us not popular science writing but high school textbooks. Simplifying, but not simple: successful popular science writing requires robust extrapolation and imagination of the kind that, I will argue, drives archaic mythologies. Some popular science writers allude to such an ambition. In a treatise that embraces the Enlightenment vision of a triumphal, encompassing science, E.O. Wilson insists that this science will give us a "new mythos" (1998:265).[13] Martin Eger (1993:206,n.6) notes, "many scientists object to Wilson's, and others', use of the word *myth* in this way." Another way of achieving similar effect is to do as Stephen Hawking does, and merely allude to "the mind of God" as the telos of science.[14] Whether explicit or coy, this is what popular science writing shares with myth: the inclination to go beyond mere findings and, through a myriad of extrapolative devices, to venture into visions that aspire to grandeur and something like wisdom – the inclination, in other words, to mythologize.

THE CRITIQUE OF CENTRIC THINKING: THREE MOMENTS

In the remainder of this chapter I expand on a point mentioned above, specifically the antiquity of the problem of myth and anthropocentrically biased knowledge. Although this background is not strictly necessary in order to follow the arguments presented in the chapters that follow, a bit of historical background can greatly enrich one's understanding of the mission accepted by popular science writers.

Students today learn early in their careers to beware of "-centric" thinking, whether anthropocentric, Eurocentric, or ethnocentric. Anything ending in "centric," they hear, will skew their analysis, leading to factual distortion and perhaps moral oppression. Anthropocentrism is one – and historically, the first – framing of the epistemological problem of centric bias. As such, concern with anthropocentrism links the critical framework of contemporary popular science writing to the very origins of Western analytical thought. Rather than attempting a fuller intellectual history, I offer here sketches of three defining moments, each of which has contributed to the backdrop of contemporary popular science writing.[15] These moments concern the development of the ideas of myth and anthropocentric illusion in relation to science and popular science writing, not the history of the genre of popular science writing (an eminently worthy topic, but not one that I pursue in depth in this book).

First Moment: Impiety toward the Gods

Although Xenophanes is even today occasionally cited as primordial parent who showed us the problem of epistemological centrism (for example, by Carl Sagan [1994:23]), Copernicus has largely replaced him as the hero of this achievement. This shift between heroes is significant – Copernicus signalling a revolution not only in understanding of the physical cosmos but also in the way that centric bias is conceptualized. In the comments that follow, therefore, I will emphasize the ways in which Xenophanes' perspective differs from contemporary understandings as much as the ways in which it gave rise to them. I begin by repeating the famous fragment from the presocratic philosopher Xenophanes: "The Ethiopians say that their gods are snub-nosed and black, the Thracians that theirs have light blue eyes and red hair. But if cattle and horses or lions had hands, or were able to draw with their hands and do the works that men can do, horses would draw the forms of the gods like horses, and cattle like cattle, and they would make their bodies such as they each had themselves" (in Kirk et al. 1983:169). This fragment alludes, tantalizingly, to centric inclinations of several kinds. For example, we encounter what looks like a strategy sometimes attributed to modern cultural anthropology: to reach an understanding of the self by means of a detour through the "other."[16] Specifically, Xenophanes recognizes *ethno*centrism or, more precisely, a kind of race-centrism: peoples of different physical type correspondingly portray their gods differently. Yet it is not this ethnocentrism or race-centrism, but rather what we now call anthropocentrism (or, more specifically, anthropomorphism), that Xenophanes picks out as the problem worthy of attention. A *diversity* of ethnic expressions of divinity is brought in merely to demonstrate that a single process is operating: the shaping of gods in human image. Thus, at least according to the tradition that credits Xenophanes as initiator of the topic, it is specifically *anthropo*centrism that goes onto the intellectual stage first, as the original focus for the problem of centric bias in the construction of knowledge.

Xenophanes also anthropomorphizes cows and other animals, noting that they would boomorphize, hippomorphize, and so on if they, like us, were to create pantheons. But the fact that his own anthropomorphizing of animals escapes comment by Xenophanes – in other words, that Xenophanes does not seem to regard anthropomorphizing of animals as part of the same process (and problem) as the anthropomorphizing of divinity – forces us to ask just how similar the problem that Xenophanes understood is to the problem as we now see it. While there are elements of continuity, it is also clear that Xenophanes' formulations rest on an image of the cosmos that is no longer the favourite of Western academic cosmology. In Xenophanes and his early descendants we encounter a cosmic absoluteness and stability that are theological in tone and inspiration. The cosmic image is something like the "Chain of Being," or a ladder on which humans occupy a lower rung than divinities. The problem is specifically the incapacity of the *lower* to comprehend the *higher*. Non-human animal species, by definition lower than the human species, do not present to human understanding the same kind of problem that gods do.

But if the lions and horses are not intrinsic to the problem being posed, why does Xenophanes bring them in? We can only speculate. Among other things, Xenophanes' anthropomorphizing of lions and horses hints at the genre-logic of the fable (interestingly, the legendary figure of Aesop is sometimes placed in the same period as Xenophanes). For whatever psychological reason, basic moral lessons gain intensity and memorableness when projected onto other species, as in the case of talking animals or even radically transformed versions of ourselves such as miniaturized humans.[17] This point should be kept in mind when, in the following chapters, we encounter Lilliputians and a miniature fire-maker (chapter 2), bacteria laughing scornfully at human arrogance (chapter 3), and other fabling gestures by popular science writers. My focus is mythologizing, but nearly all traditional and originally oral folkloric genres, including legend, proverb, anecdote, epic, and travellers' tales, have been tapped by popular science writers, as if those who take up the mission of enlightening the masses feel impelled to speak in the forms they imagine to be those of their audience. The results range from clever to strained.

As an alternative to anthropomorphic deities, Xenophanes advocated a higher notion of divinity: gods that are unchanging, self-sufficient, and free of human caprice and desire. Plato followed along similar lines, although what troubled him was less the "ethnic diversity" visible in portrayals of divinity than the moral vulgarity of the gods' behaviours as depicted in mythological tales (especially as recounted by Homer and Hesiod) and the impiety toward the gods that such humanizations implied. "I mean the greatest and most malevolent lies about matters of the greatest concern: what Hesiod said Uranus did to his son Cronos and how Cronos revenged himself on his father. Then there is the tale of Cronos's further doings and how he suffered in his turn at the hands of his own son, Zeus. Even were these stories true, they ought not to be told indiscriminately to young and thoughtless persons. It would be best if they could be buried in silence" (*Republic* II.378 [1985:74]).

As his alternative to the vulgar deities of traditional mythology, Plato advocated what might be termed philosophers' deities – notions of divinity inspired by the perfection and immutability of any true Form and intuitable through the rigorous dialectical argument that was the province of the philosopher. The unremarked irony, of course, is that like the gods of the Ethiopians and the horses, the gods of the philosophers bear a distinct resemblance to those who proffer the portrayals; the effect is particularly striking in Plato's student Aristotle, from whose account one could gather that gods spend their time doing philosophy.[18]

The intellectual problem of anthropomorphizing divinity made its way, through Augustine among others, into Christian theology, where it still has some resonance. While one encounters echoes of Xenophanes' problem in a few contemporary popular science writers, notably Paul Davies, who speculates on the convergence of ideas of modern physics with theological ideas, most contemporary popular science writers are influenced more by the problem as it took shape in the "second moment" (see below).

However, an important legacy remains from the time of this first philosophical confrontation of anthropocentric bias. Specifically, this early confrontation gave rise to a dichotomy between allegedly lower and higher forms of knowledge: the former more story-like, familiar, earthy, and entertaining; the latter, more lofty, abstract, skeptical, and reputedly analytically rigorous. The former caters to our self-delusions, giving us what we want to hear; the latter leads us out of delusion and toward objective truth. But why mince words? One is right, the other wrong. Something like these valences are what popular science writers have in mind when they invoke myth as a foil for science.

Second Moment: Ignorance, Dogged Tradition, and Enlightenment

In the seventeenth and eighteenth centuries, the era of the Enlightenment, the championing of science and reason over dogged tradition and superstition produced important shifts in the conceptualization of anthropocentric bias. Most importantly, deductive philosophizing took second place to empirical observation in pointing the way to knowledge. Correspondingly, the root of mythology was seen to lie not merely in the intellectual slovenliness that Plato decried but also, if not more so, in the *simple empirical ignorance* that must have marked the condition of humans in their initial state – the state in which mythology was born. Scenarios of such an initial state were popular among the Enlightenment philosophes and were fed from two sources: first, a general infatuation with the idea of composing a universal history of mankind; and second, the rapid growth of ethnographic knowledge that would make such a history possible, for prior states of humanity were still observable, so it was thought, among the "savage" peoples of the world. A heroic story of science steadily lifting humanity out of ignorance would form a unifying thread or metanarrative.

In the early eighteenth century, Bernard Fontenelle (*Of the Origin of Fables*), for example, speculated about such an archaic state, and the mythology it produced:

> This philosophy of the first ages worked on a principle so natural that even today our philosophy has no other; that is, we explain the unknown things in nature in terms of those we have before our eyes, and we carry over into natural philosophy ideas supplied us by experience ... These poor savages, who were the first to live in the world ... could only explain the effects of nature by grosser and more palpable things with which they were familiar. And have we not done the same? We are always representing the unknown by analogies with what is known to us, but, happily, there is all the reason in the world to believe that the unknown cannot but resemble what we presently know.
>
> From this crude philosophy, which necessarily ruled in the early ages, were born the gods and goddesses. It is quite curious

to see how the human imagination has given birth to false divinities. (1972:12)

Fontenelle's argument both distances the age of myth and connects that age with our allegedly more enlightened era. The *distancing* inheres in his asserting the falseness of those archaic, divinely centred theories of nature, attributed in turn to a rudimentary state of knowledge. Things unknown, at that rudimentary time, were accounted for by projecting onto them familiar forms of causality, notably that of human agency, and thus did the gods originate. But Fontenelle also *connects* that era with ours by positing common mental operations that propel both the earlier failed natural philosophy and our more successful ventures. Operating in ignorance, those early savages did their best; but we now know better. For Fontenelle and his contemporaries, the idea that the gods were created by humans to explain nature – by projecting familiar human agency onto it – could not help but raise skepticism about religious explanations of nature and the cosmos in general. And thus, anthropomorphism, which for Xenophanes was a problem *within* religion, became increasingly, for many in Fontenelle's era, the problem *with* religion.[19]

But the philosophes of the Enlightenment also recognized that ignorance could not be the sole explanation of mythology, for traditional, parochial, superstitious accounts of natural phenomena continued to exert a blind appeal even in the face of scientific advance. The attraction exerted by traditional explanations, and a seeming human vulnerability to them, was a source of frequent wonder, as well as exasperation, among the philosophes (and it is a theme yet encountered in contemporary popular science writers, especially the more polemical of them, such as Richard Dawkins[20]). These linked themes – ignorance giving rise to mythology as a first attempt to understand nature; the gradual displacement of ignorance with true knowledge; and the continuing, seemingly intractable appeal of dogged superstition – have all become part of the accreting heroic story of science and its nemesis myth. This second moment of theorizing about anthropocentric bias continues to dominate the conception of myth that one encounters in contemporary popular science writing.

Poised between the heroic story of science born of the Enlightenment and clearly modern themes born of psychoanalysis, Freud a bit later created what is now regarded as the classic statement of science

confronting, and in a series of steps transcending, the anthropocentric visions of myth (or of religion – for purposes of the present
argument there is no great distinction[21]):

> In the course of centuries the *naive* self-love of men has had to
> submit to two major blows at the hands of science. The first
> was when they learnt that our earth was not the centre of the
> universe but only a tiny fragment of a cosmic system of scarcely
> imaginable vastness. This is associated in our minds with the
> name of Copernicus, though something similar had already been
> asserted by Alexandrian science. The second blow fell when bio-
> logical research destroyed man's supposedly privileged place
> in creation and proved his descent from the animal kingdom
> and his ineradicable animal nature. This revaluation has been
> accomplished in our own days by Darwin, Wallace and their
> predecessors, though not without the most violent contempor-
> ary opposition. But human megalomania will have suffered its
> third and most wounding blow from the psychological research
> of the present time which seeks to prove to the ego that it is not
> even master in its own house, but must content itself with scanty
> information of what is going on unconsciously in its mind.
> (1964:284–5)

Freud's success in capturing a protean vision of science is attested by
the many inventive variations that popular science writers still continue to work on this triadic formula, with Copernicus at the base
(these variations are explored in chapter 6).

Copernicus forms the base of this by-now familiar triad, and presently one can find no hero more revered for clearing the ground for
science. Copernicus compiled observations and calculations to support a basic correction to our understanding of the cosmos, but his
present-day symbolic significance goes far beyond this correction.
The Copernican Revolution has come to stand as emblem par excellence of revolutions that are particularly radical, in which we come
to see that not just one or a few arguments, but an entire system
of understanding, is precisely backward. The Copernican Revolution also stands for revolutions that are particularly profound in the
sense of having deep emotional and/or moral repercussions – revolutions that shake the very foundations of moral and cultural life,
as Copernicus is often said to have done in regard to the religious

worldview of his time. Finally, the Copernican Revolution is invoked to epitomize the depth and tenacity of our anthropocentric inclinations and, more broadly, the tenacity of religious tradition in the face of scientific observation and evidence.

All of these associations have made the Copernican Revolution a favourite symbol of contemporary popular science writers. But one further implication is sometimes drawn from Copernicus, this one more problematic (and likely beyond anything Copernicus himself would sanction). Merging with other notions of "relativity" in the air in the twentieth century, Copernicus for some has come to represent the first step toward the recognition of the reversibility of any relationship – the *necessarily perspectival* character of *any* observation or vantage. With this implication, the problem of anthropocentrism severs its tie to the context of cosmic absoluteness within which it originated. Having gone on record first with Xenophanes' conjectures about the hierarchical relation of gods and humans, the problem of centric bias now exists as well in secular versions and is seen by many as having opened the door to relativism in its many forms. The idea of the perspectival relativity of *any* point of view also sets the stage for the population boom in "-centrisms" that we have witnessed in the last few decades.

Third Moment: The Family of Centrisms

A third development belongs to the twentieth century, especially the latter part. In the milieu of cultural relativism and postmodernism, a population boom of "-centrisms" has occurred, as though the original critical insight has spread and simultaneously subdivided. Anthropo-'s children now include, among others, Ethno-, Euro-, Phallo-, Logo-, Biblio-, Metro-, and a variety of ad hoc creations that, yes, includes Science-.[22] A bastard-child is Hodie-centrism (literally, Today-centrism), joining the spatial metaphor of a centre with a category – time – often set in ontological contrast to space. Placing "Anthropo-" at the apex of the genealogy does not reflect the order in which humans invented various centric frames of thinking and behaving (let's assume they all began long, long ago), but rather its founding place in the history of systematic thinking about epistemological bias. A deep skepticism toward the idea of "disinterested" knowledge currently pervades the academic mood in the humanities and certain segments of the social sciences, and the population boom of centrisms will not abate until this skepticism lifts.

Unlike the first two moments described above, which contributed directly to the frame and rhetoric of contemporary popular science writing, this third moment has presented some major challenges. When epistemological bias could be portrayed as all of one type, science could be held up as a uniform antidote. By contrast, the recent proliferation of centrisms injects a vision of humanity splintered into different interest groups (males vs. females; literate vs. non-literate; city vs. rural, and so on) each possessing a distinct, constitutive slant on the world. The seeming inevitability and robust diversity of bias threaten the promise of universal betterment depicted in the myth vs. science contrast of the optimistic second moment. The response from popular science writers has been mixed. Stephen Jay Gould, for example, seems to be intrigued by the variety of forms that bias can take, although whether he sees his own is another matter. Steven Weinberg, by contrast, has acquired something of a reputation as defender of the possibility of pure science in the face of the charge of science as shaped by systematic (e.g., male) bias.

Focused on epistemological bias, the present book partakes of the spirit that gave rise to the population boom of centrisms. But rather than attempting to advance a fashionable new one, I reconsider the original centrism – the most general, human-species level one – within a dynamic and expanding object, that of contemporary popular science writing. My focus on anthropocentrism arises partly in response to the importance of this concept within the long-term history of the study of myth and of science. But it also stems from the fact that anthropocentrism, in part because of its antiquity as a problem, can stand for the problem of centric bias in all its varieties. And there is this quirk: in the contemporary splintering into a range of centrisms, with *anthropo-* dwindling to one form of centric thinking among many, the critique of centric thinking, viewed broadly and historically, becomes *more anthropocentric*. Instead of worrying about our construction of the gods and other things beyond the human, we have become absorbed in analyzing our constructions of ourselves.

Still, within this obsession, numerous levels of centric inclination can operate at once, and I do not mean to close off analysis of any of them; indeed the analyses I present show little respect for borders. In the passage which opened this chapter, Xenophanes, in trying to tell us about what we now call anthropocentrism, spills over into ethnocentrism (Ethiopians vs. Thracians) and into various forms of zoocentrism (the gods as non-human animals would have them). Xenophanes thus serves as founding symbol not only of anthropocentrism as a problem but also of the bleed-over and structural commonality between different forms of centrically construed knowledge claims.

This third moment presents dilemmas not only for popular science writers but also for mythologists. In the first two moments characterized above, anthropocentrism and mythology were inseparable because the critique focused on portrayals of human-like gods. But what about this third moment, in which gods are not a primary concern? For many theorists, "myth" continues to be tied to the idea of centric bias even after deities have been largely dismissed, as if centric bias and the strategies it feeds are more critical to the concept of myth than are deities. "Myth" is increasingly invoked in most any context to designate knowledge claims thought to be slanted toward the interests of one particular human constituency

I offered Xenophanes and Plato as exemplars of the first moment of the critique of centric thinking, and Fontenelle as exemplar of the second. As an exemplar of the third moment, I offer Roland Barthes' modern classic *Mythologies* (1995), which deals with the bourgeoisie worldview and its ideological mechanisms and imagery. The concept of myth elaborated by Barthes does not require gods: bourgeoisie myth is a "naturalizing" discourse that locates the reason why things are as they are, and finds assurance for a Eurocentric vision of the world not in the whims of a pantheon, but in processes equally beyond human control. "Naturalizing" is the attempt to portray social conditions created through human volition as though they are rooted in nature and thus beyond human culpability.[23] Although "naturalizing" and "anthropomorphizing" superficially might seem to be opposites, they are equally anthropocentric and precisely complementary: the latter term designates our attempts to claim responsibility and cosmic influence where we in fact have none, while the former designates a modern version of a traditional mythic strategy – that of scapegoating[24] – aimed at removing responsibilities that we do have but don't want, by projecting them as forces lying outside of ourselves. The idea of a "nature" that has been liberated from the gods thus holds out a new sort of Cosmic Absolute and court of last appeal, one as subject to mythologizing as the world of supernaturals.

And yet, in reading Barthes, one cannot help but note the number of allusions, or in some cases secular parallels, to doings of the gods in ancient mythologies. In Barthes' vignettes we encounter beings who achieve a kind of technological apotheosis (see the "Jet-man" [1995:71ff.]) as well as contemporary humans described as ancient or timeless (see "The Face of Garbo" and "The Brain of Einstein" [1995:56ff,68ff.]). Also, we have the origin of the cosmos; for example, Jules Verne "built a kind of self-sufficient cosmogony, which has its own categories, its own time, space, fulfillment and even existential principle" (1995:65). In the especially popular "'Blue Blood' Cruise," Barthes reflects on the current-day journalistic fascination with catching European royalty wearing short sleeves and print dresses, pretending to be "mere mortals" (1995:32). Here one might recall the sexual encounters between gods and humans in Greek mythology, dalliances which confirm the divide between these realms as much as any relation between them. The new mythology subtly exudes an ancient feel, serving up ancient themes in modern and nominally secular dress. The concept of myth elaborated by Barthes is less a theoretical overturning of Xenophanes, Plato, and Fontenelle, than an updating of them.[25]

While I have emphasized the interrelation of the concepts of myth and anthropocentrism, it will become evident in the following chapters that a third idea also frequently intertwines with these: the dualism of spirit and matter, or, in its modern Cartesian form, of mind and body. Such dualism, some insist, itself functions anthropocentrically, serving to legitimize the idea that humans are above the rest of nature. Once again the idea of myth and the critical stance toward it that one encounters in Barthes mark an updating rather than an overturning. For the ancient religious conception of higher spirit and lower matter metamorphoses into modern, secular strategies of *intra-human* oppression, whether in social class, race, gender, or politics. One repeatedly encounters portrayals of the allegedly higher forms (men, Europeans, Caucasians) as possessed more of mind; and of the allegedly lower forms (women, non-Europeans, people of colour) as possessed more of the body's needs and limitations. The powers to exalt and oppress contained in the ancient duality of matter and spirit survives even in myths that arise in a world that is unsure about spirit, eliciting a critical response that, while updated, reflects and solidifies the ambivalence toward myth that took root millennia ago.

This book is about the contemporary florescence of a particular genre, that of popular science writing. But in a broader sense it is about a particular chapter in the history of an issue central to Western critical thought – one that, at least as much as any other contender, deserves to be called *the issue* of Western critical thought – namely, centric bias, an epistemological defect that long ago became associated, if not synonymous, with the idea of myth. The three moments considered above present us with three theories of the source of mythic bias. In briefest terms: human moral/intellectual slovenliness and self-infatuation; empirical ignorance leading to the projecting of the familiar on the unknown; and will-to-power exerted as images of the cosmos that promote hegemonic cultural/political regimes. The diversity of alleged motives merely confirms the tenacity of the association of myth with centric bias.

In view of the fact that the critics of myth considered here have come to be seen as pivotal just because of the harshness of their judgments, it would be well to keep in mind two moderating considerations. The first is that it is not the glaring, but rather the more subtle, implications of these critiques that should concern us. One does not have to be convinced that mythology or popular science writing is "slovenly" to recognize that both of these engage many different forms of appeal, some more intellectually rigorous than others. Without imputing ignorance, we might consider the degree to which traditional myths or contemporary popular science writers are willing to pronounce cosmic lessons for humanity in conditions of highly incomplete empirical knowledge. Will-to-power is a strong term, but hierarchy in many forms is a root condition of human life (and maybe life in general); and humans elaborate hierarchy in complex ways that often involve forms of denial. Concerns and anxieties about hierarchy – whether social or cosmic (i.e., our status in the cosmos) – are often tapped, either blatantly or subtly (even subliminally), both in traditional mythology and contemporary popular science writing.

The other consideration to be kept in mind is that the criticisms of myth summarized above represent the negative pole of a set of attitudes toward myth whose overall tone is ambivalence. The shrill critiques have always been accompanied by an equally insistent fascination with myth's aesthetic appeal, psychological and sociological insight, and moral and political power. What I want to demonstrate is that popular science writers are engaging in the same sorts of argument and persuasion that they criticize in myth. But on the larger, surrounding questions – is myth a good thing or bad thing, and is it a condition we can or wish to try to transcend? – I am not attempting to offer closure. My views on such matters no doubt reflect the ambivalence of the longer intellectual tradition of which they are a part.

☙

The segment of the longer history of myth's alleged bias to be considered in this book is, of necessity, full of invention, for popular science writing answers to antithetical imperatives. The cosmos that popular science writers strain to mythologize is not the cosmos of archaic imagination at the dawn of humanity, but rather the cosmos of science, a cosmos composed of "matter," a neutral, impersonal substance born of the desire to *demythologize*. Hence the contradictory impulses in popular science writing: on one hand, to leave myth behind; on the other, to strive toward its power, scope, and appeal. In the struggle between these impulses we catch a unique glimpse of what humans "really want" from the cosmos. As alluded to above, such writing does not constitute the first instance in which myth has been the object of contradictory desires. The tension is not entirely coterminous with the "modern" moment that Bruno Latour fruitfully problematizes in his *We Have Never Been Modern* (1993); rather, it is much older. In *How Philosophers Saved Myths* (2004), Luc Brisson describes the process by which, in the wake of an early inclination among some classical Greek philosophers to ostracize myth, it was instead "saved" for posterity, through allegorical reinterpretation, by the same intellectual movement that spawned the detractors. Who better to sense the need for the annihilated than the annihilator? Between that moment and our own one can find both significant similarities and significant differences. Regarding the differences, imagine a continuum of ways of saving myth. At one end lies a cultural situation in which a corpus of specific, broadly known stories, ripe for allegorizing, forms the coin of the realm for aspiring thinkers. At the other end is a felt need to offer something "of mythic proportions," but operating without an acknowledged canon of stories; thinkers grope with fragments, vague memories, and elemental symbols (light vs. dark, birth vs. death, freedom vs. bondage, etc.), concocting a nebulous brew that saves "myth" more than "myths." Although in both moments, antiquity and the present, aspects of both ways of saving myth(s) are to be found, antiquity on the whole would seem to lie more toward the first modality, our moment more toward the second. This suggestion is offered provisionally – merely as a possible first step in attempting to interconnect the question of where we are now with that of where we have been in our ongoing negotiation of the place of myth in human life.

2

It Had to Be You!
Fire without Prometheus

True, there are many savage tribes and some civilized peoples who tell stories of a time when their ancestors were without fire, and who profess to relate how their forefathers first became acquainted with the use of fire and with the mode of eliciting it from wood or stones. But it is very unlikely that these narratives embody any real recollection of the events which they profess to record; more probably they are mere guesses invented by men in the infancy of thought to solve a problem which would naturally obtrude itself on their attention as soon as they began to reflect on the origin of human life and society. In short, most if not all such tales are apparently myths. Yet even as myths they deserve to be studied.

Sir James Frazer, *Myths of the Origin of Fire*

We thus begin to understand the truly essential place occupied by cooking in native thought: not only does cooking mark the transition from nature to culture, but through it and by means of it, the human state can be defined with all its attributes.

Claude Lévi-Strauss, *The Raw and the Cooked*

We ignore the impossible at our peril.

Michio Kaku, *Physics of the Impossible*

Scientists do not lay great evolutionary transitions on the shoulders of heroes from mythology;[1] yet dispatching mythic heroes is an easier task than dispatching mythic thinking. In *The Artful Universe* (1995; expanded edition 2005[2]), John Barrow, professor of mathematical sciences at the University of Cambridge and Fellow of the Royal Society, takes the reader on a tour of the cosmos, one that traverses scales from microscopic to astronomic. The tour is designed to reveal the operation of laws (or "constants") of nature within realms that many regard as quintessential sites of human freedom and spontaneity: consciousness, culture, and art. More broadly, Barrow argues that in long-running debates about heredity vs. upbringing in the determination of human character – the so-called nature vs. nurture controversy – public opinion has moved too far in the direction of nurture; balance must be restored! In the midst of Barrow's complex, urbane rumination, which cycles with ease between galaxy, atom, and art museum, sits a scenario of the origin of culture through the domestication of fire. Fire is the image of all that is, or appears to be, free, spontaneous, expansive. We became human by conquering it, while as symbol of spirit or mind it became the pre-eminent trophy of our species. The scenario of domesticating fire brings to a crescendo Barrow's scientific vision, which developed in the context of debates about the so-called "anthropic principle" – the controversial idea (existing in many versions and degrees of radicalness) that the universe is calibrated to humans (or something like humans), at least to the extent that the universe had to be a certain way to allow the emergence of observers.[3] In the analysis that follows I pursue Barrow in his more recent course, dealing not with the anthropic principle generally but rather with its engendering concerns as they are drawn, as if by underlying affinity, toward a scientifically naive doppelganger in the world of myth, namely, the story of fire and its role in gaining for our species a special cosmic place.

The most powerful myths are those that go unrecognized as such; I propose to show that, tweaked to scientific ends and shorn of a named hero, what Barrow gives us in his fiery crescendo is none other than a new version of the ancient story that Western readers associate with Prometheus, although in fact this story exists in numerous variations of plot and hero through much of the world. Aimed at a modern audience, Barrow's analysis takes the outward form of science, an enterprise often rhetorically opposed to myth. My claim is that once we peel away the dazzling but thin scientific veneer from Barrow's scenario, we are left with a colourful, ingenious, "just-so" speculation with deep affinities to, and clearly belonging among, those assembled by James Frazer in his now-classic study, *Myths of the Origin of Fire*.

SCIENCE MEETS MYTH

Since I add a mythologist's comparative framework to Barrow's already expansive project, some background discussion is in order. I begin by considering, under the images of the butterfly and the wall, the ideas of chaos and "sensitive dependence" – themes that have a kind of parallel existence in the universes of science and mythology. Then, with King Kong as exemplar, I consider a commonplace strategy of science popularization that Barrow, in a much elaborated version, has put at the centre of his analysis; this strategy rests on juxtaposing science with popular fantasy of a type often labelled as mythology. Thirdly, I discuss Barrow's ideas about the underlying relationship of science and art and the notion of constraint that informs these. All three of these concerns contribute to the unusually evocative cultural ambience of Barrow's project, one in which myth and science bump up against and flow into one another. Following these brief ruminations, I turn squarely to Barrow's pronouncements about the origin of culture through fire.

The Butterfly and the Wall: Chaos and Sensitive Dependence in Myth and Science

Recent decades have seen a widespread popular interest in systems described by scientists variously as "complex," "non-linear," or "chaotic." The last of these terms, "chaotic," has special resonance for mythologists, calling up images of unformed or disordered primordial conditions depicted in many creation stories.[4] Chaos in the scientific context calls attention to states defined by both determinacy and unpredictability, while in the context of myth, chaos tends to imply the most radical antithesis of order and of deterministic principle. Despite this divergence, however, one finds points of contact between mythic and scientific chaos. Scientific chaos theory, for example, calls attention to "sensitive dependence on initial conditions," the so-called "butterfly effect." The latter term invokes a hypothetical scenario in which, through the temporal ramifying of consequences, a butterfly flapping its wings produces or alters the course of a hurricane, and thus deals a major blow to weather prediction. To a mythologist the scenario feels not entirely unfamiliar. Air (or more broadly, weather) is insubstantial yet capable of bringing fortune or ruin; whimsical yet not entirely devoid of pattern, the weather's defiance of human attempts at domination forms a worldwide mythological theme – the rebellious Typhoeus and his descendants in Greek mythology offering only one of many variants.[5] But even more abstractly, the butterfly effect has resonance for a mythologist, for the piddling cause that ramifies into an enormous consequence forms the plot of numerous mythological tales. Many myths, for example, portray the origin of human mortality as a consequence dramatically out of proportion to its cause: death originates for humans because a messenger from the gods confuses the message that he is to deliver, or because of some other stupid, trifling error that occurred in primordial times.[6]

Whatever the overlap between mythic and scientific chaos, however, it does not extend to the language employed to describe chaotic states, for scientific chaos theory invariably resorts to mathematical models that are not part of the arsenal of traditional mythologies. To be sure, traditional mythologies frequently do ruminate on quantitative concerns: the magnitudes and shapes of things, the periodicities of nature, and numerological patterns of many kinds.[7] But traditional mythologies do not typically make use of complex mathematical formulas, Cartesian graphs, or logarithmic notation; these fall within the province, and indeed serve as emblems, of science. I recall a serious public television documentary on chaos theory in which, to the accompaniment of a numinous ambient sound, the audience was asked to *gaze upon the very shape of chaos*: what appeared on the screen was not a photographic image of cloud formation or whirlpool, but a computer-generated "strange attractor," the mathematical graph that represents chaotic states. For me, as no doubt for most viewers, the mathematics was partially grasped at best. But the graph, whose form also reminds some of a butterfly, was an attractor in a second sense as well: of viewers to public television and to the cause of science. Beyond its power (which I do not doubt) to tell us something about the physical world, the abstract mathematical form holds another kind of power, more iconic or even totemic; this other, evocative power is one of the concerns of this chapter.[8]

∞

Along with the "butterfly," this chapter also focuses on another image, that of a "wall." In the usage that concerns us, a "wall" is a quantitatively defined threshold beyond which a given effect cannot occur – for example, the relation of parameters limiting the size of living things that can make effective use of surface tension to walk on water, as some insects are able to do. The wall sometimes emerges as a sort of response and/or complement to the butterfly. If the butterfly delicately initiates a mad propagation of consequence, the wall offers to contain the madness; where one cannot predict what will happen, one might seek compensation in a science dedicated to predicting at least what will not. To counter the challenge by "sensitive dependence" to the idea of a predictable cosmos, an ambitious analyst might even strive to establish a series of intersecting walls that together winnow out all possible courses of events save one, thus in a roundabout way bringing back the classic idea of a predictable and necessary cosmos.

This chapter focuses on an analytical framework developed by an astronomer, John Barrow, while the next chapter focuses on an analytical framework developed by a paleontologist, the late Stephen Jay Gould. Amidst many differences, Gould's and Barrow's analyses converge around a dialectic of butterfly and wall: both scientists grapple with the blow to the ideal of predictability struck by "sensitive dependence" and "contingency" in evolutionary process; and (although only Gould actually uses the term) both rebound by invoking something like "walls" to underpin their respective visions of regularity within the cosmos.

King Kong in a Universe of Impossibilities

Science popularizers not infrequently adopt as a strategy the demonstration of the real world impossibility of a familiar character or phenomenon drawn from the realm of fantasy – a gimmick or "hook" motivated not so much by skeptical zealotry as by a concern to make science fun. Such demonstrations inevitably invoke specific walls that demarcate the realm of the possible from the impossible. Approaching King Kong from a structural engineering point of view, for example, brings into focus a number of such walls. In enlarging a three-dimensional object, the ratio of surface area to mass declines, and at a certain scale King Kong, without design modifications, would overheat. The skeleton would also be inadequate for the mass, and there would be many other problems. Cleverly presented, such analyses allow the pleasure of fantasy to metamorphose into the pleasure of science.

In this chapter I examine a key scenario from John Barrow's recent book, *The Artful Universe*, specifically Barrow's arguments as to why a very small creature could not have domesticated fire. Prefacing his arguments with a review of giganticist and miniaturist fantasies (Gulliver's Lilliputians, giant insects of 1950s B-movies, and other "impossibilities of structural engineering" [1995:64]), Barrow clearly locates his arguments within the engineering King Kong genre of science popularization.[9] However, even though similar in principle, Barrow's analysis of the fire-maker differs vastly from the King Kong example in scope and complexity. Specifically, Barrow's arguments about the fire-maker constitute merely the culminating moment of a grand attempt to delimit the possible from the impossible on a cosmic scale. As we will see, Barrow pushes "to the wall" the possibilities of wall reasoning.

The search for walls is particularly fascinating when carried on in the context of cosmic evolution. Barrow and many other thinkers have addressed the question of whether our knowledge of the minimum physical requirements for the evolution of life and consciousness on this planet – and hence, of conditions that would have ruled these out – allows us to draw conclusions about the possibility, impossibility, or degree of likelihood of life or consciousness in other regions of the universe or in other universes. Barrow's *Artful Universe* provides a particularly robust rumination on such issues, and thus stands as particularly revealing of the character and the limitations that arise in endeavours of this general type.[10] In terms of the *overall* topic of my book – mythic anthropocentrism in its subtler forms – my interest in this particular current of thought can be summarized tersely: I argue that deployed in the context of cosmic evolution, wall thinking opens the way to sly new forms of anthropocentric invention.

Science and the Art of Cosmic Constraint

In *The Artful Universe*, Barrow presents a vision of the human place in the cosmos, one dominated by the constraints on human physical and mental makeup imposed by the constants of nature. Some of these constraints belong to matter in general; for example, our human body form and size are subject to the same gravitational and thermodynamic laws that constrain all matter. Other constraints, while rooted in matter in general, are specific to human psychology; for example, humans, according to Barrow, are universally predisposed to artistic representations of landscape that reflect the savannah conditions of our biological evolution. In the title of his book, Barrow attaches the modifier "artful" to "universe" rather than to the human species; his intention is less to anthropomorphize the cosmos than to naturalize art by challenging the idea that human aesthetic sensibility exists as an autonomous realm that can be understood apart from the rest of the natural world.

As his broadest concern Barrow engages, especially in his Introduction and Conclusion, the relationship between the sciences and the humanities. The recent past, he argues, has witnessed a large gap between the sciences and the humanities in approaching human nature. Science has traditionally pursued the goal of generalization and prediction. The humanities (with which Barrow tends to link the social sciences) have resisted these goals, emphasizing instead the variable, creative, and plastic character of human behaviour, especially in the realm of aesthetics.

> Anthropologists and social scientists have traditionally laid great stress upon the diversity of human artistic and social activity, but largely ignored the common features of existence that derive from the universality of our cosmic environment, and the necessary features that life-supporting environments must display. Just as science has for too long focused almost exclusively upon the regularities and simplicities of the world at the expense of the irregularities and complexities, so our contemplation of the arts has over-indulged the diversities and unpredictabilities of its forms at the expense of the skein of shared features that bind us with these forms of complexity to the underlying environment that the Universe provides. (1995:viii; cf. 245–6)

Barrow, like many other popular science writers, thinks that the present moment holds out the possibility of convergence between science and the humanities. Recent work on "complexity," he claims, has led scientists to recognize chance and contingency alongside uniformity and predictability; in this respect, science has moved in the direction of the humanities. *The Artful Universe* offers Barrow's case for why the humanities should return the compliment: by acknowledging the many regularities that operate in art, the humanities can move, reciprocally, in the direction of the physical sciences.

Barrow's invitation is accompanied by a thoroughly art-friendly tone. Unassumingly he presents himself as a man of art as well as science, interspersing his reflections on evolutionary psychology with long discussions of various schools of Western and non-Western art. Virtually all of his chapter and subchapter titles allude to literary or other artistic works, suggesting that the framing devices of art can also frame the presentation of science. The chapter that contains the fire scenario (chapter 3 – Size, life, and landscape), for example, also contains subtitles with these phrases: "of mice and men," "jagged edge," "war and peace," "far from the madding crowd," and "*les liaisons dangereuses*."

Although Barrow himself does not make a point of it, *The Artful Universe* also holds out an interesting challenge to those social scientists and humanities scholars who assume that modern cosmological science supports the doctrine of cultural relativism. Scholars in the humanities in particular, typically with meager knowledge of the specifics, have sometimes glibly invoked "general relativity" or "quantum uncertainty" to suggest a kind of cosmological backdrop for the idea of cultural relativism.[11] Barrow's analysis provides a counter to this tradition, for he offers a modern, scientifically grounded cosmological treatise that aims not to buttress, but rather to rein in, what he sees as over-inflated notions of aesthetic and cultural plasticity. Interestingly, too, food customs figure pivotally in some of Barrow's arguments; intended or not, this focus also calls to mind a longer anthropological tradition of cultural anti-relativists repeatedly going to battle with cultural relativists specifically over food customs. On one side are scholars of a materialist slant who seek to account for food tastes and customs through a universal cost/benefit calculus, sometimes with protein or calories as the currency. On the other side are scholars who offer the variability of food tastes and taboos as the epitomizing example of cultural variability and plasticity – an anthropological inflection of the venerable adage that there's no accounting for taste (more on this debate follows).

∽

It is important to note that Barrow's full perspective emphasizes principles of both contingency and necessity in cosmic evolution. Invoking recent insights into "complexity," he periodically elaborates on the *chance* nature of cosmic evolution: things might have gone differently at many points.[12] But Barrow's greater interest lies at the pole of necessity. Even in a contingent universe, he argues, we find pervasive regularity based on constraint: however contingent the course of development, at the end of the day the results will lie within specifiable walls, within realms of possibility defined by constants of nature. Barrow defines such realms of possibility by plotting intersecting variables of mass, size, and force (e.g., gravitation, surface tension); and he does so at several levels. For example, on the broadest level, all physical bodies from atoms to galaxies exist within a specifiable ratio of size to mass; above that ratio they fly apart, below it they implode and turn into black holes or fall into a "quantum region" that excludes any observation. For organic life to exist, a number of additional conditions must be met, such as those that specify the ranges within which a planet can sustain an atmosphere.[13] Moving from matter and life to the evolution of consciousness and culture, we encounter yet further constraints. For example, a being must have a certain minimum physical size if it is to have the capacity to read books, namely, the size below which pages would be unmanageable because they would stick together or to the being's hands (i.e., the adhesive forces would be too great relative to the being's page-turning musculature). Although Barrow presents one range of possibility (defined by the intersection of two variables) at a time, he overlays various ranges one on another so that the effects are cumulative. A book-reading being must first fall into a certain ratio of size to mass in order to be a physical body that neither flies apart nor implodes; then inhabit a planet within the parameters that allow for an atmosphere that supports the evolution of life and brain; and finally lie outside the size range in which page-adhesion prevents it from handling a book. Of course these are just some of the ranges that can be explored in visualizing the cosmic possibility of a book-reading being.

Most of the ranges of possibility that Barrow proffers rest on the principle that structures or designs that work at one scale will not necessarily work at another.[14] As noted earlier, many readers will already be familiar with such arguments, since they have become a favourite of popular science presentations in recent decades. However, compared with most examples, including the elementary King Kong example considered earlier, the multiplication of levels that we find in Barrow – the attempt to specify the range within the range until we reach an entity as specific as a book-reading being – adds enormous methodological complexity. The attempt to use walls to compounding effect deserves careful attention, particularly given the role (which Barrow clearly acknowledges) of contingency within cosmic evolution.

Let us now turn to Barrow's arguments about the domestication of fire.

ECCE HOMO: THE FIRE-MAKER WINDOW OF POSSIBILITY

Barrow's chapter 3 (Size, life, and landscape) is the most ambitious section of *The Artful Universe*. Here, in mapping out regions of cosmic possibility and impossibility, Barrow proceeds from general and encompassing to local and specific, starting with the fundamental parameters that constrain any physical body (from galaxies and stars to insects and atoms), moving on to the requirements for biological life, and culminating in the physical requirements for a specific form of life, that of a conscious, culture-bearing being. To put it another way, Barrow attempts to locate the walls that divide the possible from the impossible at three junctures typically recognized in the organization of matter, namely, the inorganic, the organic, and the "mental." The philosopher Henri Bergson referred to these as the three great leaps in nature: from nothing to something, from something to life, from life to mind. Barrow's discussion includes graphs that portray narrow lines of existing, possible things, surrounded by great blank spaces that depict the realm of the impossible. The first of these graphs, which serves to convey the flavour of all of them, corresponds to his broadest claim: that whatever the size and mass of an object, the ratio of size to mass (i.e., the density) remains constant; that the entities of the cosmos form a "line of constant density" (1995:51). This "line" is evident in Figure 2.1.

Other graphs and analyses follow this one. Then Barrow attempts to demarcate the subdivision of organic life within which rapid mental/cultural evolution is possible; his arguments at this point give prominence to the hearth, for Barrow sees the domestication of fire as stimulus for the sort of rapid mental and cultural evolution that has characterized our species. It is as though an abstract, analytical wall converges with a concrete wall, the one that demarcates the firepit. Barrow offers a scenario of the domestication of fire – one which I will not hesitate to call a variant, and an inspired one at that,

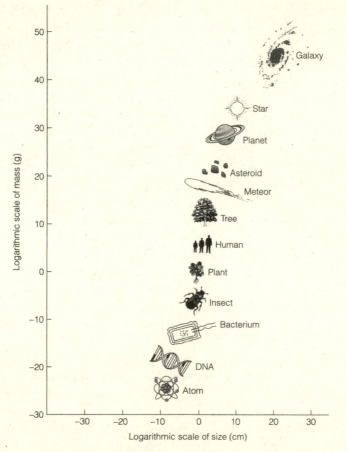

Figure 2.1   The masses and sizes of some of the most significant structures found in the Universe. Illustration, Figure 3.1, from p. 49 of *The Artful Universe*, by John D. Barrow (1995). By permission of Oxford University Press.

of the traditional myth of the origin of culture through the domestication of fire. His analysis and diagrams convey a sense of the complex balance of forces required for life and mind to exist, with the implication that if there is to be a conscious being in the cosmos it has to be approximately us ("approximately" will become clearer as we proceed). Anything else would be walled out as a suitable fire-domesticator. His method based on ruling-out thus turns into a method of ruling-in.

Let us consider the specifics of Barrow's account of the domestication of fire, which begins with the physics of a successful fire:

Our ability to use fire is ... connected with our size. There is a
smallest possible flame that burns in air, because the surface area
surrounding a volume of burning material determines the influx
of oxygen that can sustain combustion ... It is a nice coincidence
that coal, wood, or peat fires have to be of a minimum size in
order to maintain the ignition temperature of the ingredients
under typical atmospheric conditions; and that minimum size
is just about what is required to keep a human being warm in a
natural shelter of convenient size.

These considerations place a restriction upon how small one
can be and still make use of fire. If there were no limit on how
small a flame could burn in air, then very small creatures could
use fire to supply warmth, initiate technologies, and change their
environments. But because there is a smallest flame, very small
creatures are faced with approaching unmanageably large fires if
they are to sustain them with fuel. Their inability to control fire
is not only crucial in preventing them from developing various
forms of technology; it also restricts their diversity. They cannot
spread into regions where climatic fluctuations are large; they
cannot populate regions where the mean temperature is very low;
and their activities are restricted to the hours of daylight if their
light-gathering sensors respond only to visible light. (1995:65–6)

I begin my critique of Barrow's account with a word of justification regarding the very idea of such a critique, for Barrow's tone is
sometimes speculative and, hence, I might be accused of taking his
fire scenario in the wrong spirit. For example, the concepts of possibility and impossibility, which run through many of Barrow's writings, have found their way into his argumentative syntax. Regarding
social and technological development, Barrow says: "it is even possible to argue that our size has been the most important enabling
factor" (1995:65). How does one respond? Perhaps I should say, "it
is also even possible to argue that this claim is far from proven"; but
what do we gain in shifting to the subjunctive?[15] Admittedly there
can be a fine line between, on one hand, proving the validity of a
particular perspective and, on the other, speculating about the ways
in which that perspective, assuming it to be valid, might be useful
in accounting for various specific circumstances that fall within its
purview. I do not object to speculation provided that speculative
claims, and more to the point, the inferences that derive from them,

are acknowledged to be speculative. It must be noted, however, that the tentative tone fails to appear in the opening and closing pages of Barrow's book, where the politico-academic lesson of the cosmos is announced: that social sciences and humanities have overestimated the variability and plasticity of human behaviour and aesthetics. If I appear overly intent on Barrow's ruminations, it is because Barrow is overly ready to draw from them grand cosmic lessons.

The idea of the plasticity or spontaneity of human behaviour has long been associated with the symbol of fire and its mastery; this symbol provides one of the means through which Barrow constructs his lesson. As already noted, the myth of the human acquisition of fire for cooking – and with this, the possibility of culture – is the focus of one of the world's most widespread origin myths. The story tells how we humans, through the deed of the Greek Prometheus, Polynesian Maui, Native American Coyote, or another such hero, gained a tool that allowed us a certain freedom to expand and design life as we would have it – a measure of autonomy and self-directedness vis-à-vis the whims of nature and the gods, though sometimes at the cost of their wrath.

Brief mention of an important scholarly study will help to flesh out these matters. Beginning with mythic stories collected from South American Indian societies, spiralling outward to other continents, anthropologist Claude Lévi-Strauss, in his work *The Raw and the Cooked,* argues convincingly that traditional mythological stories are often structured around correlation of oppositions that in different domains contrast life in a state of nature with life in a state of culture; these oppositions include nudity vs. dress, isolation vs. exchange (including endogamous vs. exogamous marriage), rudimentary subsistence vs. subsistence aided by tools and technology, ill-mannered vs. well-mannered social behaviour, and, pre-eminently, eating raw vs. cooked food – the latter often mediated by a colourful story of how fire was originally obtained.

Typical of the many mythic narratives assembled by Lévi-Strauss is one belonging to the Kayapo people of South America; it tells of a jaguar that adopts a human boy and shows him for the first time a bow and arrow and the knowledge of how to hunt and cook meat. The boy ends up fleeing the jaguar's home and returning to his human village to tell of his adventures. On hearing these, the villagers proceed to the jaguar's home and succeed in stealing the bow and the fire. Lévi-Strauss' rendition of the story concludes: "For the very first time it was possible to have light in the village at night, to eat cooked meat, and to warm oneself at a hearth. But the jaguar, incensed by the ingratitude of his adopted son, who had stolen fire and the secret of the bow and arrow, was to remain full of hatred for all living creatures, especially human beings. Now only the reflection of fire could be seen in its eyes. It used its fangs for hunting and ate its meat raw, having solemnly renounced grilled meat" (1969:67; original text from Banner 1957).

This particular story emphasizes the correlation between the acquisition of technology (the bow and arrow) and the transition from raw to cooked food; in other versions of the story, acquisition of cooked food correlates with development of other qualities enumerated above (exogamy and exchange, clothing, well-mannered sociality, etc.) that distinguish a cultured existence from life in a "state of nature." This Kayapo story reflects several patterns broadly characteristic of Native American mythological narratives, including a predilection for plots involving abduction or adoption of humans by animals, and an inclination to attribute to animals capacities and resources that humans admire and need (of the kind that in other parts of the world are often attributed to gods).

But even at great geographic distances one finds myths with similarities to the Kayapo story; for example, the Greek story of Prometheus also contains the motifs of acquisition of fire by theft, together with the wrath of those from whom the fire is stolen. Of Prometheus' theft of fire the Greek poet Hesiod tells us, "This bit deeper into the heart of Zeus the thunder-god: he was enraged when he saw mankind enjoying the radiant light of fire. In return for the theft of fire he instantly produced a curse to plague mankind" (507–616 [1953:69]).

Turning to Barrow's fire rumination, one also encounters resonances with these stories, including a tendency, which we explore shortly, to think of fire, cooked food, and tools/technology as parts of a single endowment. And although the motif of theft itself is absent from Barrow (at least in any obvious sense), there is another, curious preconception bound up with the theft motif in traditional fire myths that cannot be so easily dismissed in the case of Barrow, and that should be kept in mind in the following analysis. Specifically, in traditional mythologies fire is sometimes portrayed as a substance that can be possessed only by a single earthly species, a conceit sometimes mythically expressed in a "zero-sum" accounting for its distribution. In the Kayapo story, for example, such an accounting is displayed in the jaguar renouncing fire once humans have stolen it; in other versions of this myth one encounters explicit contests over possession of this substance, one species' gain being the other's loss. While it could have arisen easily from the human experience of being exclusive controllers of this substance, such a conceit stands in marked contrast to the nature of fire itself as a phenomenon available in virtually limitless supply.

∽

In focusing on fire and its place in the acquisition of culture, Barrow, intentionally or not, thus has chosen a particularly appropriate mythic target – our myth of freedom – for the project of revealing hidden constraint. Barrow's arguments – or perhaps the Olympian gods still seeking revenge against the wayward Prometheus – hold out the possibility of a perverse counter-project: that of defending the message of our myth by challenging the idea that the myth is necessarily *ours*. I fall into that perverse project willy-nilly, for, although I respond as readily as anyone to mastery of fire as mythopoeic symbol of all that is human, Barrow's arguments fail to convince me, in the scientifically literal way in which he intends them, that the fire-domesticator had to be us (or approximately us).[16] Barrow also implies, in arguments still to be considered, that the domestication of fire figured pivotally in the origin of culture; once again, I am unconvinced that *literally* this had to be so. The traditional mythological poetics of fire suggest a parallelism in which the spontaneous, active character of fire (as microcosm) resonates with the expansive, creative character of culture (as macrocosm). While such themes clearly run through Barrow's analysis, he sometimes also taps the parallelism of fire and culture with the opposite intention, i.e., to reinforce parallel *limiting* claims concerning micro- and macrocosm: as fire itself is constrained (a flame must have a certain minimum size to support combustion), so also is culture constrained (beginning with the requirement, asserted by Barrow, that to develop it a being must have a certain minimum size imposed by the physics of fire). One does not have to object to the possibility of constraints operating on any or all of these levels to find fault with the particular arguments about constraint that Barrow has put forward; these grow increasingly difficult to prove or disprove, or even decipher as arguments, as one moves from fire as substance to the requirements for the fire-maker and ultimately to the nature of culture.

Barrow's focus on the *size* of the fire-maker might seem a novel angle to bring to a theory about the origin of culture, but in fact this consideration too is old, for size also figures importantly in some traditional fire myths. To take, once again, the example close to home: Prometheus, descended from Titans, is typically character-ized as a "large" being, instancing another common mythic conceit, that of correlating physical dimensions with importance, such that gods and primordial ancestors are typically portrayed as larger than humans of today. Barrow follows his fire arguments with a discus-sion of several disadvantages of very large size (1995:70ff.), thus in a general way dampening the quest for a fire-maker even larger than us. But the arguments he presents that deal specifically with fire offer comparisons only to beings smaller than us, with no correspond-ing attempts to rule out beings larger than us specifically as fire-makers.[17] Thus, within the range of sizes that Barrow actually con-siders for the fire-maker, *Homo sapiens* emerges as the Titan. In both the ancient story of Prometheus and Barrow's modern re-casting, it is a big guy who gets control of fire. Contests between beings of small vs. large stature are a favourite theme in myths and other forms of folk narrative; and, briefly glimpsing forward, I call attention to the fact that this theme, too, along with the earlier-mentioned employ-ment of wall arguments, will connect this chapter with the follow-ing one on Stephen Jay Gould. Impressive formulas and calculations abound in both Barrow and Gould; yet on one level their opposed conclusions about the human place in the cosmos come down to alternative outcomes to the story of David and Goliath or the fable of the lion and the mouse. Both Barrow and Gould cast humans as a physically large species pitted against a smaller one (the hypothetical small fire-maker in Barrow, bacteria in Gould). In Barrow's account the big species wins, whereas Gould's explicit purpose is to cosmic-ally dethrone the human species; accordingly he will tell us how the large is brought down by the small.

෨෧

To assess Barrow's claims, I will first analyze the formal character of his arguments in relation to the evidence he presents; following this, I will move on to rhetorical considerations, looking at extra-evidential factors that may lend persuasiveness to these arguments.

In terms of the formal character of his analysis, I see one persistent problem: Barrow continually proves less than he claims to prove. The discrepancy between claimed and proven is evident in two main ways. First, as part of the process of narrowing down the range of possible sizes for the fire-domesticator, Barrow restricts his purview in ways that are *anthropocentric* and sometimes *ethnocentric*. His analysis, nominally aimed at demonstrating the unworkability of a "small creature" as fire-maker, really only considers a miniature version of himself or his likely reader – a Western *Homo bourgeois*. He does not consider the possibility of a small being with a physical structure or a culture radically different from his own, and thus ends up proving only that a small being could not be Barrow. The second shortcoming of Barrow's proof belongs to the cost-benefit ratio computations that figure prominently in his analysis. Barrow's study lacks precision; we end up with evolutionary "just so" stories with a veneer of quantitative effect provided by very general graphs and frequent references to vaguely quantitative factors. In the following comments, I expand on these two points, moving back and forth between them.

As noted above, Barrow's investigation of possible fire-makers focuses mainly on ruling out "small creatures." But Barrow seems to take it as a given that the small being would occupy what might be termed a studio-shelter (he notes that the minimum size of a coal or peat fire "is just about what is required to keep a human being warm in a natural shelter of convenient size" [1995:66]). This coincidence sits well with Barrow's discontent with cultural relativism, since the preference for single-individual- or even small-group-size dwellings is, to say the least, not universal among humans. Let us change this variable, which at base is not even a physical factor but merely the cultural preference of some societies. As an alternative we might entertain the idea of a *tribe* of small beings who stoke the fire. None of Barrow's data allows us to conclude that one of these scenarios overall is more efficient or evolutionarily advantageous than the other, so we cannot rule out the possibility that what actually happened in our evolutionary history is merely a contingent outcome. In the absence of proof, there remains rhetoric, and one could easily pull together a case in favour of small beings who manage to take advantage of centralized heating; the nice coincidence that the smallest peat fire can sustain one human could just as easily be the nice coincidence that it can sustain a tribe of small creatures. The nagging, deeply seductive, fatal flaw peculiar to arguments about the *possible* is prejudice in favour of the *actual*.

Barrow's discussion of the fire-domesticator is both preceded and followed by commentaries in the same vein regarding other capabilities that are typically singled out as milestones in the development of culture, including books, use of the wheel, and use of tools large enough for hunting and for shaping rock, wood, and metal (1995:65–7). Barrow shows that we are the right size for all of these. For example, "to make use of materials like paper, it is necessary to be large" (1995:67); the adhesive forces would make paper and books unusable for small beings, who could not turn the pages. The same forces would render small wheels unusable. Thus, alongside the claim that we are the right size for making fire, Barrow presents other benefits of our large size, including the fact that we are the right size for many tools and for making use of paper and wheels. He also loosely connects our size with the possibility of bipedalism and its importance for the development of manual dexterity and mobility (1995:65). Barrow is not precise about the parameters of the culture-bearing being. And yet: large like us, bipedal, uses wheels, reads books – does this remind you of any species you already know?

෨෧

Barrow is intensely interested in the evolution of culture and in theoretical issues surrounding the concept of culture, yet major equivocations run through his use of this concept. The most significant of these arises from inconsistencies in the degree of abstractness with which he elaborates the concept of culture; Barrow ends up conflating culture in the abstract with culture in the concrete. Abstractly, culture refers to any way of life organized by ideas, values, symbols, and information (as opposed to genetic encoding or instinct); when Barrow poses the big questions about the evolution of a conscious being in this universe, he seems to be thinking of culture in this abstract way. By contrast, culture invoked concretely by Barrow is books, wheels, cooked food, and weapons. The concept of culture in this concrete usage is radically narrowed, coming to designate one possible set of technical means for, and manifestations of, life organized through ideas, values, symbols, and information. These means and manifestations look suspiciously like the technological milestones of our own species-history, or rather, the milestones we have been taught to recognize as emblems of the human "rise to civilization" (amongst which, however, one finds little about the indigenous technologies of South America or Polynesia, for example).

An inquiry into the origin of culture – as opposed to one particular culture – would leave *open* the issue of means and manifestations, and recognize at least the possibility that a smaller species might develop concrete means through which to accomplish what wheels and books accomplished for our species. When posing questions about conditions for the possibility of the evolution of culture, any intrusion of the concrete notion of culture into the abstract notion introduces a circularity that functions anthropocentrically: the more we infuse the concept of culture with the specific, concrete details of our own species-history, the more we will be forced to conclude that it took a being like us to achieve culture. We arrive at the astounding conclusion that we are the right size to wield the particular technologies that our species invented. It is not by cosmic constraint but by definition that the cultured/conscious being will have to be like us. Attempting to lay out the prerequisites for a being capable of consciousness and cultural evolution is a difficult, groping business – one greatly aided by hindsight. The nature of consciousness is one of the great topics of our era, and so far its precise nature has proven resistant to unifying interpretation (some of the contentious issues appear in chapters 4 and 5). We are still far from consensus about what the thing is whose minimal requirements Barrow is trying to deduce.

As noted above, Barrow invokes the abstract notion of culture when talking about the significance of culture within evolutionary process: "By passing on ideas through social interaction, by means of language, records, images, symbols, gestures, and sounds, our development has proceeded far more rapidly than by encoding particular types of information in genes. The information that can be passed on by these behavioural and cultural means is of a type that cannot be transferred by genetic inheritance" (1995:82). This definition of culture is not very different from those offered by cultural anthropologists, who along with others, according to Barrow, have gotten carried away with nurture over nature. Note that nothing in the above definition requires books, wheels, or any particular technology; for that matter, nothing in the definition requires any technology at all, since even very elaborate records can be committed to memory, as demonstrated by countless oral historians and genealogists in non-literate human societies. The lack of any or all of these technologies does not of itself make any way of life less cultural than any other way of life that is organized by ideas, symbols, and information. However, it might appear to do so from the perspective of parochial, ethnocentric equations that most of us at times draw, for example, between culture and books. Note the openness inherent in the concept of culture quoted above; indeed the definition rests on a contrast to that which is genetically *fixed*. It is this openness that inspires cultural anthropologists to speak of cultures rather than culture;[18] it is just this openness that disappears when Barrow turns from abstract theorizing on the nature and significance of culture to specific arguments that attempt to deduce its physical and organic prerequisites. At this point we have a silent narrowing of culture to Barrow's culture.

೦๑

Barrow is not wrong, of course, to point out that small changes, at any level – in the size of Earth, its orbit, or its atmospheric composition, for example – might rule out life and consciousness as we know them; but we have to consider "sensitivity" in the other direction: the possibility that alternative forms of life or manifestations of consciousness might be opened up by such changes. If one accepts contingency, one cannot conclude that all that is possible has happened. If we are to prove that "a small creature" (not just a small us) could not have domesticated fire, and we seriously admit the idea of contingency in cosmic evolution, we must throw open the possibility of a small creature in the sense of *any* small creature – allowing the possibility of a structure different from ours – and we must throw open this possibility on the level of the fundamental matter of the universe in all of its possible permutations through the time span of the evolution of the universe.

Alternative scenarios compound, and grow more fundamental, as we move back in time. Contingent events open some possibilities and close down others; as we move in temporal regress we face the task of identifying and rescuing for consideration possibilities that have been contingently forestalled. Can consciousness arise only out of biological life as we know it? Must life be based on carbon? Can consciousness occur in a universe with less than three dimensions?[19] Could the constants themselves vary throughout a "multiverse"?[20] We encounter the dread butterfly flapping in reverse as we proceed into the past.[21] Barrow as much as admits that evolutionary process finds ways around the sort of walls that form his stock-in-trade; he notes, for example, that "[a] reason for the ubiquity of fractal designs in Nature is that they offer a general recipe for escaping the strait-jacket on design that is imposed by the simple relation between volume and surface area that we find in regular objects" (1995:59). That natural processes can give rise to "engineering solutions" that if designed by humans we would call brilliant, and which human understanding must catch up to, does not make the job of rescuing cosmic possibilities easier.

THINGS THAT DO NOT HAPPEN

Thinkers have often felt a mysterious power in negative concepts: negative logical operators, negative numbers, the "void," "taboo," "absence," "nothingness," "negative dialectics," and so on. An elaborate philosophical analysis of the type of negative claim Barrow is proffering might be very enlightening, and metaphysics of some sort is unavoidable here. But I suggest that insight into the character of negative prediction might also be derived from a project opposite in spirit to metaphysics, specifically the eminently *practical* business of "risk management." This business is not without metaphysical awareness: the very existence of risk management is predicated on secular, naturalistic formulas that nevertheless insist on referring to "acts of God." Yet risk management differs from metaphysics in directing its energy toward surviving within the metaphysical condition of contingency rather than speculating about it out of purely intellectual intrigue.

The part of risk management dedicated to the humble business of "home security" suggests a number of parallels to Barrow's analysis. Barrow's fire-maker parameters have the effect of keeping evolutionary intruders out of our hearth, and, as noted above, many fire origin myths (including that of Prometheus) portray the acquisition of fire as an act of theft; already we seem to be on the right track.[22] More importantly, the home-security business, like Barrow's analysis, aims to discover conditions for the possibility of non-events – the parameters that guarantee that a certain thing will *not* happen (though often, in the case of home security, settling for an acceptable probability that it will not). In either case, the complexity of the analytical task is "sensitively dependent" upon the degree of generality with which the goal is articulated: correctly installing a deadbolt on a house full of windows has about the same significance for home security as ruling out a bourgeois homunculus domesticating fire has for ruling out "a small creature" domesticating fire. In each case, just one potential violation, the most obvious, is eliminated.

Rigour takes on a special character in the context of arguments directed toward ruling-out. It takes a thief to catch a thief: the thief turned to defeating thievery engages in an internal dialectic, alternatively proposing and attempting to defeat particular imagined scenarios. Though imagination is often held to be antithetical to rigour, arguments directed toward ruling-out form the glaring exception: in this case, rigour *demands* imagination. Interestingly, Barrow himself argues that art and science have a common origin in imagination (1995:5). However, while *The Artful Universe* is a creative and highly imaginative work, one cannot help but notice how little of Barrow's imaginative power is directed toward exploring the possibility of alternative evolutionary scenarios or alternative fire-makers.[23] The requirements of an empirical proof of Barrow's claims are far from clear. I am not aware of any definitive evidence at this time suggesting any other instances of life in the cosmos, let alone any other fire-makers; indeed the idea of Earth's exceptionalism has been taken up in a smattering of recent works (see chapter 6). But even if in all of the universe and through all cosmic time elapsed thus far there were found no fire-maker nor conscious being radically different from us, would the sum of all of these absences prove Barrow's claim? He seems to want to show not why there *is* not, but rather, why there *cannot* be, another fire-maker. This is a strong reading of Barrow's claim, but one at which a reasonable reader could easily arrive.[24]

Moreover, the consequences of reading Barrow's claim less strongly are no less dire for his proposed lesson about the doctrine of cultural plasticity and variability (specifically his claim that that this doctrine should be reined in). One might choose to read Barrow, for example, under the assumption that he is merely showing that constraints of various kinds operate at many different levels of organization that have no fixed relationship; and thus that one should not see in his expository progression, from matter to life to mind and culture, a cumulative narrowing-down of the range of possibilities from one level to the next. But among other things, such a reading would leave open the possibility that a minor contingent change at any level could open new possibilities at supervening levels, thus opening the way for "solutions" to any isolate obstacle. Indeed, by whatever route one would read Barrow's claims less strongly, to that extent the claims fall prey to another weakness: they are increasingly trivial and non-predictive because they are more easily gotten around, especially as one approaches the realm of the book's ultimate lesson, i.e., cultural plasticity and variability.

Arguably a contingent universe would always contain (so to speak) things that *could have happened but did not* just because they chanced not to and/or because other contingent events forestalled them. As a form of evidence, absence gets shakier in proportion to the robustness of the idea of contingency. Proving a negative has always been a methodological hobgoblin; proving negatives within the context of contexts – cosmic evolution – is certainly no less so.

CONSTRUCTING EFFECTIVE FIRE-WALLS: ON THE IMPORTANCE OF PRECISE MEASUREMENT

Barrow asserts that "any structure that we see in the Universe results from a balance between opposing forces of Nature" (1995:50); his opening examples have to do with ratios of mass and size that hold for all matter. As Barrow, in an almost seamless transition, moves from the gross constituent matter of the universe to biological and cultural evolution, concern with the balance between size and mass is replaced by a focus on trade-offs between costs and benefits.

Regarding the contributions of fire to human survival Barrow says this:

> Its principal benefit is the possibility of having a barbecue. Cooking makes food easier to consume and digest, kills harmful bacteria, and enables meat to be preserved for longer. These gastronomic factors serve to enlarge the range of foodstuffs available to fire-making humans, improve their health, and reduce the range over which they need to search for palatable prey. Cooking also stimulates the emergence of a discriminatory sense of taste.

Meat can be cooked in a variety of ways; its taste differs from
that of uncooked meat. The nuances of taste that cooking creates
and the division of labor that it entails have clearly played a continuing role in human social evolution. Only hominids practiced
cooking, and, unlike other animals, we take trouble to make
food look nice as well as taste good. (1995:66)

Consider the generalities in this statement. Quantitative in aspir-
ation and effect ("easier," "longer," "enlarge the range," "reduce,"
etc.), there are *no* specific measures or explicit cost-benefit calculations (and very little is added in the context that surrounds this
passage). The distinction that Barrow has focused upon nominally
has to do with cooking, but his real concern seems to be barbecuing
meat, another prized custom of *Homo bourgeois*.

Barrow's argument can be historically contextualized as the latest,
and most ambitious, of a series of anthropological treatises that
assert a significant relation between culture and protein. It is characteristic of such treatises to challenge cultural relativism (with its
emphasis on plasticity of cultural form and on nurture over nature)
through arguments aimed at demonstrating the universal validity
of cost-benefit analysis (figured in protein, calories, or other such
measures). Not only does Barrow's analysis fit squarely within this
tradition of meaty challenges to cultural relativism but also, with
the forces of the universe converging in the barbecue, it is fair to
say that in Barrow the tradition has reached its cosmic culmination.
It is true that previous instances of such debates have sometimes
involved a cosmological dimension; for example, in *Pigs for the
Ancestors*, a foundational work in this genre, anthropologist Roy
Rappaport analyzes a Maring New Guinea ritual that is characterized by massive butchering and consumption of pork. Rappaport
says that the ritual operates as a "transducer" and "homeostat" that
optimizes human energy and protein capture (1968:233ff), but also
reports that the indigenous interpretation holds that this ritual has
to do with the relationship of humans to spirits, especially ancestral
spirits.[25] Similarly, Marvin Harris' (e.g., 1974) classic analyses were
directed toward revealing the protein-logic behind seemingly non-
utilitarian religious practices, such as Hindu taboos relating to cattle.
Here it is worth mentioning that the mythic hero Prometheus, along
with stealing fire, also stole meat from the gods for humans; specifically, in a division of an ox between gods and humans, he tricked

Zeus into choosing the bones (hence humans offer these to the gods in sacrifice), while humans retained the meat (see Hesiod 507–616 [1953:69]). In sum, one can find a variety of traditional myths and religious worldviews that cosmically enshrine particular dietary customs, including some that emphasize, even apotheosize, eating of flesh; and one can also find scientists who see a convergence between mythico-religious and scientific rationality in such customs. But Barrow shows no visible interest in such parallels or convergences: taken at face value, the cosmos that enshrines meat-eating for him is solely that of contemporary Western evolutionary science.[26]

The last few decades have produced numerous contending theories, and broad public concern, about food and diet and their place in human life. In this context, it is surprising that Barrow apparently sees no necessity to cite evidence in support of his views on such matters – as though he regards his claims to be intuitively obvious or self-evident. Notably, Barrow does not substantiate any of his arguments about the benefits of meat-eating; the complexity of the issue would increase greatly when the benefits are weighed against costs of procuring meat. Considered in context, the claim that cooking "stimulates the emergence of a discriminatory sense of taste" suggests that there is a link between cooking and broader powers and/ or forms of aesthetic/cultural discrimination. Yet the only evidence offered supporting such a link occurs in a terse footnote, which happens to contain Barrow's only reference to the elaborate traditional mythology of raw vs. cooked food: "We still identify cooking and eating habits as a form of social distinction. This is not a recent innovation: the Algonquin Indians of north-east America disparaged their northerly neighbours by calling them 'raw meat eaters', that is 'Eskimos'" (1995:66). Barrow uses this bit of ethnographic data to buttress only the idea of a link between cooking and social distinction. He could not have pursued the matter much further without confronting the fact that most of the energizing themes in his own vision of fire in the development of culture are already to be found in the ancient, venerable mythology of fire – a circumstance that raises interesting questions concerning the genesis of Barrow's "scientific" ideas.

From the standpoint of sensation, taste is largely a matter of smell, a capacity for which humans are less distinguished than many non-cooking species. From the standpoint of learned (or cultural) aesthetics, the question is not whether cooking fosters powers of discrimination, but whether it does so more than any other expressive medium that might be developed by us or an alternate culture-bearing species. In his comments on the emergence of discriminating taste, Barrow once again conflates the abstract notion of culture with something else: this time, with a notion of culture that, while overlapping with the "concrete" notion of culture discussed earlier, emphasizes the artistic more than the technological – specifically, a normative or "epicurean" usage in which a "cultured" person is a person of discriminating "taste." It remains to be seen whether Barrow's implied causal link between gustatory and aesthetic/cultural taste more broadly is a physiological and bio-evolutionary fact or merely a convenient though powerful and widespread cultural trope, one capable of engendering elaborate, evocative mythological scenarios.[27] A final comment on the concept of culture: Barrow's views on cooking in human evolution also hint at the so-called leisure theory of culture. The passage from Barrow (on pages 57–8) opens with an account of the practical nutritional benefits of cooking, but then deviates without a rationale into behaviours with less obvious utilitarian benefit. The latter are summarized in the claim that with the emergence of cooking "we take trouble to make food look nice as well as taste good" (1995:66).[28] Here too Barrow's scenario suggests a Promethean legacy, for leisure and culture are sometimes bundled together in the poetic notion that by bequeathing to us this magical tool, source of energy, and means for lengthening the daylight, Prometheus gave us for the first time the leisure to pursue philosophy, art, science, or other activities of less immediate utility.

The very idea of a connection between mental/cultural evolution and the domestication of fire must raise suspicions of a mythic intrusion into Barrow's science. And whether one credits myth or science as Barrow's source, myth wins: either myth is a mode of understanding powerful enough to worm its way into and take over Barrow's nominally scientific account of our mental evolution; or else, if Barrow's account really is science, myth must be credited with having deciphered the formative event in the evolution of culture millennia before Barrow was able to do so scientifically.

THE CASE FOR A SMALL FIRE-MAKER

If one wanted to construct or justify a plausible, alternative course of evolution, in which small beings domesticated fire, how would

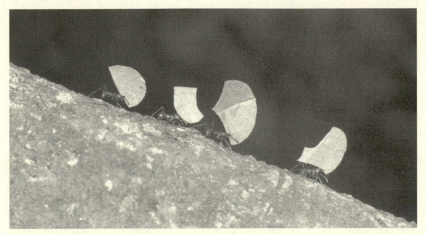

Figure 2.2   The fuel brigade. The ability of small beings to carry loads many times our own weight assured our victory in the quest to domesticate fire.

one go about it? Interestingly, one finds information useful for this task in Barrow's book. For example, in a part of his presentation on size but unconnected with his arguments specifically about fire, Barrow calls attention to the greater population sizes possible at the small body-size level (1995:68ff.). This is a factor that could facilitate the formation of a fuel-brigade.[29] At one point in Barrow's book (1995:55, Figure 3.3), there is a photograph of an ant carrying a bit of leaf that overwhelms it in size, accompanied by a caption noting the ability of ants to carry loads greater than their own body weight. History is written by the victors; consider the use that revisionists could make of this sort of photograph (Figure 2.2) if contingent evolutionary processes had taken a different turn and small beings had gained control of fire.[30]

Calculating the cost-benefit ratio of two alternatives within a single society within a single species at one point in time is complicated, necessitating consideration of complex systemic ramifications. A cost-benefit analysis of fire-making within all possible courses of cosmic evolution would require a multivariate formula of monumental proportions. But instead of attempting such a formula, Barrow resorts to a quantitative strategy of an entirely different sort, one that I will call *quantitative impressionism*.[31] This form of quantitative analysis works by blending a few basic but eye-catching graphs and other symbols of science with a smattering of phrases that vaguely connote measurement ("easier to consume," "preserved for

longer," "improve their health," "reduce the range"), and then adding humanly evocative images of a more elemental kind: juicy barbecue; warm, aromatic peat fire.[32] Please! I am already convinced: the
benefit is *definitely* worth the cost – our evolutionary ancestors took
the right course, one that could only lead to greatness in our species!

೨೦

Barrow argues that "very small creatures are faced with approaching unmanageably large fires if they are to sustain them with fuel"
(1995:66). The fact that he repeats this argument in the much briefer
rendition of his fire scenario in another work (1998:124) suggests
that Barrow views this as a particularly critical consideration. But
one can envision many solutions to this obstacle, both in terms of
body and fire design and social organization. For example, masses of
fuel could be piled up, and then a small flame or ember transferred
to it. In the extreme, some of the drones might be forced into self-
immolation, and once more the large populations and socio-centric
character of small beings would be an asset (certain small beings
such as moths are even pre-adapted, as if waiting to acquire a purpose for their self-immolating inclinations).[33] If such measures were
employed, a time might come when technology, by allowing control
of fires many times the size of the controllers (as blast furnaces are
to us now), would render self-immolation an obsolete practice. If
the descendants of these little guys were to develop a mythology, the
self-immolating ancestors would be its founding heroes.

   In analyzing the domestication of fire, Barrow is concerned with
rather gross physical considerations such as the size requirements
for tending a fire; interestingly, he does not raise the question of
whether there is a minimum brain-size requirement for becoming a
fire-domesticator. Perhaps this is because of the impossible tangle of
issues that such an inquiry would open. Or perhaps it is because such
an inquiry would go against the general drift of his views on cause
and effect in our mental evolution. Barrow embraces the view that
it was the domestication of fire that allowed the expansion of the
brain and with it the capabilities epitomized in culture and art (see
especially 1995:67); thus, the domestication of fire would have had
to take place before, and as enabling condition of, the great mental
expansion. By Barrow's own arguments we are led to conclude that
the original fire-domesticator's brain would have been "small."

If a small being cannot be a fire-maker, it is not for any reason that Barrow has given us. Moreover, in the end, the principles of "contingency" and "maximization," both of which Barrow invokes liberally, are fundamentally contradictory. Contingent means accidental and unpredictable, while maximization implies a deterministic selection between alternatives. It is possible, of course, that the two principles could be reconciled (with each operating at specifiable moments), but Barrow does not accomplish such a reconciliation, and the irresolution haunts his analysis.

MYTHIC RHETORIC

Scientific pronouncements go down more easily when they are already familiar as myth; a major source of persuasiveness in Barrow's analysis lies in the fact that it teaches us something we already believe. The feat of acquiring culture through fire became a heroic myth long ago and, among many peoples, probably for no more (nor less) reason than that the scenario is poetically evocative and socially edifying. Modern phenomenologist Gaston Bachelard was fascinated by "the secret persistence of this idolatry of fire" (1968:4). Waxing poetic, Barrow himself emphasizes the role of fire in providing a place of warmth, security, festivity, and heightened emotion (e.g., 1995:96). Myths typically cast the evocative symbols of humanity as primordial, and their coming to be as coterminous with and integral to our own coming to be as humans. Barrow's presentation – his organization, tone, and framing – adopts a rhetoric that is congenial to the mythic values of the fire scenario he relates; it is as if he is attempting to scientifically literalize an evocative mythopoeic symbol. Given the widespread, if not universal, importance of fire as symbol, one cannot but wonder what thoughts would have arisen had Barrow explored the place of fire in art along the lines in which he considered savannah landscapes.

In the following discussion, I consider Barrow's rhetoric under three foci: first, the effect created by his progression from cosmos to hearth; second, the resonances of Barrow's scheme with the classic cosmological idea of the "chain of being"; and, third, some further arguments by anthropologist Claude Lévi-Strauss on the "the raw and the cooked" in mythological perspective. All of these foci are in some sense cosmological, and it is important to note that to cosmologize – that is, to claim to derive one's conclusions from the context of all contexts – is itself a distinct rhetorical strategy. The case in point here will be Barrow's cosmologizing of his stance on cultural relativism: while this doctrine has been challenged many times over with data on human customs from comparative sociology and anthropology, Barrow mounts his attack from general principles about the structure of the universe. The following analysis aims to identify sources of persuasiveness in Barrow's arguments that lie in rhetorical strategy, in contrast to evidence.

From Cosmos to Hearth

Barrow progresses from cosmos to hearth through macrocosmic/ microcosmic analogies, a favoured conceit in traditional mythologies. As noted previously, he leads us through three realms, each more tightly delimited than the previous: we start with things that are, then move to things that are and also are living; and proceed finally to things that are, are living, and also have culture. In some respects the three realms are analogous to one another, for the progressively more stringent criteria that delimit each realm are established through the same general means, and each of the three realms emerges as a balance of opposing forces that allows some entities to exist and disallows others. Although nothing in *The Artful Universe* suggests that Barrow is holding out for a pre-Copernican model of the cosmos, his mode of presentation has a curiously pre-Copernican feel, moving in a centripetal direction, from the cosmos at large to our planet and finally to the hearth, the symbol of the pre-eminently human.

A hierarchy emerges as we proceed to the hearth: each of the three realms – matter, life, mind – adds a new set of requirements to the previous realm, so that the final realm, spatially the most localized, is also structurally most complex, rare, and delicate. But this hierarchy is simultaneously also one of progressive encompassment since each more stringently defined realm incorporates the requirements of the previous realms and adds its own.[34] Humans end up at the summit of a scale of difficulty-of-being, each step strewn with entities that did not make the last "cut," the ghosts of ruled-out things: galaxies that cannot be because they have too much mass for their size; things that cannot be alive because their planet cannot support an atmosphere; living things that cannot have elaborate culture because they cannot cook and develop discriminatory taste since their fires would be too tiny to sustain combustion. The situation is reminiscent of hierarchies of the ruling gods in many mythologies (or, for that matter, of real-life political orders): hierarchical attainment correlates with the number of people (or, in Barrow's case, analytical walls) working to topple claimants to power and position. One should note that in Barrow's scheme the final, most stringently defined concentric circle – of the conscious, cultured being – has by far the most exclusive membership. A great variety of things exist, and many different species are alive, but fire-makers seem not to be a diverse lot. The world of the possible is broad at the base and narrows until, cone-like, it comes to a point.

The repeating of a similar schema of the possible and the impossible as we proceed in turn through the three main matter, life, and mind levels might convey a sense that the more general findings stand behind and add their weight to the more particular finding; in other words, that the particular finding (about size and domestication of fire) is supported by the more general findings (about relations of size and mass in the cosmos). World mythology suggests that humans are predisposed toward macrocosmic/microcosmic analogies, and that these work bi-directionally: not only do humans like to think of the cosmos as the village, but, in the other direction, we like to see ourselves as cosmic fractals, as local instantiations of larger structures, so that, reciprocally, the village is the cosmos. Furthermore, general-to-particular progressions tend to convey a sort of transitivity of authority. The reader may sense an analogy and be inclined toward a deductive inference between the levels that Barrow considers: as planets supporting organic life are rare within the cosmos, we, as fire-making beings, must be rare within organic life. Furthermore, since each superimposed level adds further constraints, it is perfectly consistent with Barrow's overall scheme that the last cut should, so to speak, leave only one being standing.

However, any such deductive analogizing between the three main levels would be unwarranted. If we accept that whatever occurs within life must also lie within the constraints imposed by matter-in-general, and that many kinds of bodies are ruled out by the constraints of matter-in-general, we have still learned nothing about how many beings, within the realm of organic life, can become fire-makers. Rarity of any occurrence is a matter to be investigated empirically within any one level. The general-to-particular relation that Barrow presents is not of the sort that is necessary for logical deduction between them. Although there is a continuity, or analogy, in *method* between levels of analysis in Barrow's presentation, there is simply nothing of content in the findings at any level that adds support to the findings at any other level.

In sum, a vaguely centripetal order of presentation, culminating in the hearth; a hierarchy that rises with (approximately) us humans at the pinnacle; and a general-to-specific progression in which the careless reader might see the weight of the cosmos supporting specific arguments about the appropriateness of our size for the role of fire-maker: rarely do we encounter such comfort in a post-Copernican universe. Ours is the Panglossian best of all possible worlds, if only because the only possibility is by default the best possibility. The curiously pre-Copernican feel of Barrow's analysis stems in part from the centripetal direction of analysis, but also in part from the fact that Barrow is intent to read a moral/political lesson from the physical structure of the cosmos, an endeavour that sustained a major blow in Copernican de-centring.

The Chain of Being

Let us now consider a second, entirely different, rhetorical issue: the possible resonances between Barrow's perspective and a traditional idea in Western cosmology, specifically that the natural world is organized as a "chain of being." This classic cosmological theme – in briefest terms, the idea that there are no gaps in nature since all things that can be, are – is elegantly explored in Arthur Lovejoy's classic work, *The Great Chain of Being*. Lovejoy finds the roots of the tradition in Plato's notion of the world of nature as a plenitude of timeless essences.[35]

In entertaining the possibility of an alternative fire-domesticator, I find myself continually lapsing into the argument that if another fire-domesticator could have arisen, it would have by now; surely one proof that it cannot happen is that it has not happened. Whether because we are heirs to the chain of being tradition, or because the doctrine of the chain is a mere formalization of some deeper, human mental proclivity, it is difficult to shake the inclination to assume that things that can be, are – by now at least! But surely this is a pre-Copernicanism of time: the conviction that our moment marks the culmination and termination of cosmic possibility.[36]

Whatever inclinations we have to assume that our species marks the culmination of cosmic evolution, these can only abet Barrow's arguments. Although Barrow's graphs are intended to suggest only that everything that is, is possible (nothing occurs outside parameters allowed by the constants of nature), it is easy to slip into our old habits and see in the graphs something else, something more like: everything that is possible, is (in other words, the chain of being).[37] And as regards fire-domesticators, that does seem to be what Barrow thinks: the fire-domesticator window of possibility offered by the universe has been pretty much filled by now. Importantly, Barrow's graphs, *considered in themselves*, could serve with equal ease to represent either proposition: that all things that are, are possible; or its converse, that all things that are possible, are.[38] It may seem a bit strained to suggest that a reader's musings could resonate with the chain of being while reading Barrow, but one of the peculiarities of abstract, metaphysical ideas – cause, effect, necessity, contingency, possibility, impossibility – is a seeming human susceptibility to subtle reversals of reasoning within them.

Whether source or consequence of parochial thinking, the chain of being, conceiving of the cosmos as already full, deters the imagining of cosmic possibilities untried. A preference for thinking parochially may carry practical value; indeed this disposition deserves to be considered, alongside the alleged preference for savannah landscape, for the possibility that it carries evolutionary advantage – an idea that was unavailable to Enlightenment thinkers as they pilloried humanity's dogged attachment to tradition.[39] Traditional proverbs reflect a wisdom that is adjusted to the cosmos as we know it, not as it might have been. Some proverbs embody a folk wisdom about the limits of possibility and the necessity of reining in one's designs on matter: "you can't make a silk purse out of a sow's ear"; "you can't squeeze blood from a turnip." Most of the time, following such caveats and staying with the tried and true may be the advantageous course, yet nature itself is full of transformations that test such proverbial wisdom (note that the substance from which we make silk purses derives from a source at least as improbable as a sow's ear, namely, the species of moth commonly known as the silk worm). Providing sage advice for human artifice within local circumstances, it is safe to say that the object of such proverbs is not fundamental matter in the context of cosmic evolution. Vis-à-vis such proverbs, Barrow's leap in scope is so grand that his methodology cannot keep up; the graphs and the walls are impressive, yet when considered in relation to his limitless ambition, at once cosmic and cosmopolitan, Barrow's methods emerge as suspiciously parochial.

The Raw and the Cooked

The most influential and wide-ranging work on the myth of the origin of cooking fire in recent decades is French anthropologist Claude Lévi-Strauss' previously mentioned book, *The Raw and the Cooked* (1969); I return to it now for some final considerations. Juxtaposing Lévi-Strauss' anthropological investigations into the relation of fire and culture with Barrow's reveals a striking convergence but also a divergence. First, the convergence. A recurrent theme in Lévi-Strauss' approach to myth (and his anthropology more generally) is the opposition between "nature" and "culture." In his early work, *The Elementary Structures of Kinship*, Lévi-Strauss follows Freud in proposing an actual historical/evolutionary moment when our species made the transition from a state of nature to a state of culture (for Freud this was the moment of the instigation of the incest taboo).[40] But in his later work on myth, centred around the scenario of the origin of culture through cooking-fire, Lévi-Strauss moves away from the attempt to locate actual points of transition, toward a recognition of the ideological load carried by the very notion of such pivotal points. Lévi-Strauss' later position emphasizes the idea that mind and culture in the end will be shown to be governed by laws of nature more generally. In this respect, there appears to be significant common ground between Lévi-Strauss' later position and the theme of universal constraints – constraints governing human art as well as (or, rather, as part of) the natural world – in Barrow's *The Artful Universe*.[41] Furthermore, like Barrow, Lévi-Strauss, through his entire career, engages in robust scientific rhetoric; and also like Barrow, Lévi-Strauss blends ardent quantitative mystique with laughable quantitative imprecision.[42]

But, as already hinted above, we also encounter a divergence. In jettisoning the idea of the worlds of nature and of culture as unbridgeable, Lévi-Strauss becomes absorbed in the question of why humans are intent to draw a sharp distinction between these realms in the first place (whether in scientific treatise or mythic scenario). He raises the issue of the desire of humans to stand above the rest of nature, potentially with the intention of dominating it.[43] This strand in Lévi-Strauss' concept of myth is a very old one: myth is shaped by anthropocentric conceit. The divergence from Barrow, then, lies in the fact that Lévi-Strauss, in his later position, is on the lookout for anthropocentric ideology in scenarios that portray our transition to humanity. Even while attempting to show that human culture is governed by the universal constraints of nature, Barrow nevertheless presents the domestication of fire as a pivotal occurrence that allowed a sudden acceleration of mental and cultural characteristics that distinguish us from the rest of nature; and, casting a cosmos-wide net, he all but concludes that our specifications demarcate the only possible vehicle suitable for this transformation. Barrow's demonstration of constraint ends up conveying a vision of humanity that would be appreciated by many who, resisting the idea of our subordination to nature, insist that our place in the cosmos is privileged, exalted, unique. The place in the cosmos that many mythico-religious treatises have sought to secure through the idea of our standing out from nature, Barrow secures through the idea of our subordination to nature, though in such a way that we emerge as its most singular product: through the ruling out of other contenders, we are scientifically ruled-in to this pinnacle position, with the mass of the cosmos standing behind and supporting us. Domesticated fire remains the core symbol of this emergence for Barrow, who achieves a scientific inflection of what Lévi-Strauss sees as a long-standing mythic vision.

But the topic itself – the domestication of fire – is not what is most mythic in Barrow's presentation. The focus on fire and size is not difficult to connect with mythology. Besides size, however, traditional mythologies also give great attention to time and periodicity, especially in the form of stories about how the cycles of nature were established and/or adjusted, for example, by cosmic heroes who lengthen the growing season or slow the course of the sun (as does the Polynesian trickster Maui [Grey 1906:11.ff]) so that these better suit human ends. The mythologist's exploration of Barrow's science that I have presented around the fire chapter (chapter 3) of *The Artful Universe* could as easily have been directed toward chapter 4 ("The heavens and the Earth") in which Barrow, impressed by the suitability of natural periodicities to the needs of human life, elaborates on "the reasons for the seasons" (1995:117). But while either the domestication of fire or the periodicity of nature furnishes a topic full of mythic resonance, the deeper mythicality of Barrow's analysis is revealed through more abstract, epistemological considerations that transcend any specific topic. What kind of mental operation constitutes myth? Myth for Lévi-Strauss implies a form of thought that is in some ways creative and open, yet in other ways closed. The closed aspect of mythic thought is explored most clearly by Lévi-Strauss through sociologically reversed images that occur in myths of many societies. Not unlike the scientific "engineering King Kong" illustrations considered earlier, traditional myths often portray alternatives to a society's reigning forms of social organization and worldview precisely in order to show that such alternatives would not work. By revealing the untenability of alternatives, myths thus confirm the conventional wisdom of the societies in which they are told.

For example, in "The Story of Asdiwal," Lévi-Strauss finds that stories told by the Tsimshian people of the northwest coast of North America contain portrayals of residence patterns at odds with those that exist on the ground. Regarding these mythic portrayals he concludes: "Such speculations, in the last analysis, do not seek to depict what is real, but to justify the shortcomings of reality, since the extreme positions are only *imagined* in order to show that they are *untenable*" (1971:30). But such mythic demonstrations typically take the form of opening only one variable (or at least a limited range of them) at a time to speculation; the disconfirmation of imagined alternatives is a foregone conclusion since changing one key variable in a system of interrelated parts while insisting that the others remain constant will necessarily introduce disequilibrium. According to this theory of myth, the possibility of *entirely novel designs* does not come up for consideration; mythic thought is thus stacked in favour of the status quo.[44] From this angle, too, Barrow's fire rumination is mythic: it demonstrates the impossibility of an alternative scenario by changing one key variable (size) while holding the other variables – notably those of anatomical, domestic, and social morphology, not to mention the fire-maker's cultural values – constant, as though Barrow has been mesmerized by the idea of the "constants" of nature with which he opens his discourse on size and culture. Or perhaps it is ultimately *Homo bourgeois* that mesmerizes Barrow, to the point that he slips into thinking of this being as a cosmological constant.[45] The way that Barrow *thinks about* his topic is ultimately more revealing of mythic character than is the obvious mythicality of the topic itself. Quantitative data is powerless to alter the epistemological character of a myth if the numbers are employed within a process of reasoning that is at heart mythic.

Paraphrasing what Lévi-Strauss said about Freud (specifically about what Freud said about the myth of Oedipus [1967:213]), I conclude by repeating that what Barrow has given us is more a new variant of, than an alternative to, the ancient myth of the origin of cooking fire. The fire-maker's proportions remain mythic, and we have a brilliant updating – an ancient concept revivified through fresh iconography. This new rendition blends enduring, universal humanity with the humanity of our specific time and place.

Whatever will be, will be; whatever might have been, within a cosmic evolution that admits contingency, remains at large. Responding to a number of opinions on extraterrestrial intelligence, Martin Rees, in *Our Cosmic Habitat*, offers an assessment that, I suggest, is as applicable to alien fire-makers as it is to alien intelligences (indeed, if we follow Barrow these would be the same being): "For myself, I think agnosticism is the only rational stance on this issue. We don't know enough about life's origins – still less about what natural selection can and cannot do – to say whether intelligent aliens are likely or unlikely" (2001:20). Barrow's claim that more of human culture is constrained than many humanistically inclined scholars wish to admit is a timely topic, and some of his less cosmically ambitious examples are worthy of further debate. Also, I happen to agree with the argument presented by some anti-relativists that there is as much intellectual and moral danger in overestimating as there is in underestimating human variability and plasticity. But none of this helps Barrow's fire scenario or the lessons he attempts to promote through it about nature vs. nurture and the human place in the cosmos. Conceived to account for art through science, Barrow's fire scenario more satisfactorily illustrates the reverse. Beneath the graphs lies something more elemental: an ancient and poetically powerful tale, which carries Barrow along in its currents. Such is the power of myth!

3

Randomness and Life: Sobering Lessons from the Drunkard's Walk

In this chapter I consider another scholar who, toward a quite different vision, combines three themes that figured centrally in Barrow's analysis: an emphasis on quantitative reasoning, the idea of walls delimiting the possible from the impossible in cosmic evolution, and an insistence on the importance of accidentalism and contingency within that evolution (an insistence once again inspired by chaos theory). Specifically, I consider the analysis of the human place in nature presented by paleontologist and popular science writer Stephen Jay Gould in *Full House: The Spread of Excellence from Plato to Darwin*, the work that this prolific author calls his "most beloved child" (1997:1).

In *Full House*, Gould utilizes the device of the statistical curve of distribution to analyze the nature of variation within four different phenomena: cancer survival rates, baseball batting averages, the organic sizes of planktonic forams, and life on Earth considered in totality. Gould directs his analyses toward two big claims. The first is methodological: he argues that each of the four phenomena is better understood when approached as a "full house," that is, with the assumption that the phenomenon in question is comprised of a region of variation that can be diagrammed as a statistical curve of distribution. Moreover, Gould claims that when we consider the ways in which people have previously looked at these phenomena, and others as well, we discover a widespread contemporary disinclination to recognize the full house. Gould attributes this disinclination to the influence of lingering Platonism. Most people, Gould charges, are inclined to be mesmerized by one segment of a curve and "devalue or ignore variation among the individuals that constitute the full population" (1997:40; cf. 2–4, 21). Gould's second big claim

in *Full House* is cosmological and moral. Specifically, in regard to
the final example mentioned above – life on Earth considered as a
totality – Gould claims that the full house perspective – and its analytical tool, the curve – will lead to a radically revised notion of the
human place in the cosmos. The older view, still held by many, that
our species is the culminating moment of a progressive evolutionary drive, will be overthrown by the vision of our species as an accidental evolutionary sideshow.

Full House is vintage Gould, with all the enthusiasm and clever
word play ("model batter" and "modal bacter," "platonic forms"
and "planktonic forams") as well as Gould's famed contradictoriness. As if aware of the complexity of this book, Gould makes a special appeal to his readers: those who persist will be rewarded with a
resolution of issues that for some time have been troubling (1997:2).
Since my analysis of Gould, of necessity, also becomes quite complex, it seems appropriate for me to offer and strive to honour an
analogous pledge: persist and you will be rewarded with a fuller
understanding of Gould's strategies of persuasion, including his pen-
chant for saying many things and seemingly their opposites.[1]

GOULD THROWS US A CURVE

Components of the curve of distribution that Gould emphasizes are
the left wall, the right wall, the right tail, and various measures of
central tendency (mean, median, and mode). Among the measures
of central tendency, the mode – or the most frequently occurring
value – figures most prominently in Gould's analyses (Figure 3.1).
Left walls are minimum quantitative values that define particular
phenomena: for example, for professional baseball batting averages,
the left wall is the minimum average found in professional baseball;
for the size range exhibited by living things, the smallest size required
to have a living cell; for cancer survival rates, the date of diagnosis
(i.e., the point at which the sufferer becomes a survivor). Right walls
are the outer or upper limits: the highest possible batting average,
the largest size of living things, the maximum length of survival from
a form of cancer. If the left wall represents the minimum criteria of
inclusion in a set, the right wall represents a maximal extension of
the set, its upper extreme. The vertical axis represents the number
(or percentage) of members of a set occurring at any given value
along the horizontal (left-right) axis. When the right portion of the

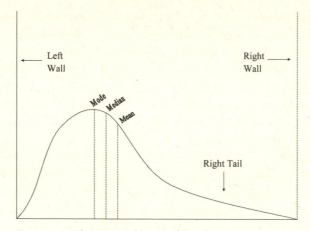

Figure 3.1   The statistical curve of distribution.

curve trails off in a long, thin extension before finally hitting the wall, the curve is said to have a right tail. The right tail turns out to be of particular concern in some of Gould's analyses, and it will be among my primary foci.

As already noted, Gould in *Full House* presents four main curves of distribution, representing, in turn, cancer survival, baseball, planktonic forams, and the totality of life on Earth (hereafter referred to as LIFE). However, cancer survival (which Gould elaborates through his personal story of surviving this disease), baseball (the national sport), and LIFE receive the main emphasis; planktonic forams are brought in unceremoniously as prefatory material for LIFE. My discussion reflects Gould's emphasis. Note that the main three amount to variations on the micro/macrocosms that, in *The Life of the Cosmos*, Smolin too (see preface) has lighted upon: self, society, and the totality of life.

As his opening example of a curve of distribution, Gould presents an inspiring account of his youthful survival, far beyond the median term, of a rare form of cancer. He attributes his fortune in part to his training in statistics, specifically to the knowledge that curves frequently have right tails. He quickly learned that such was the case with his disease; instead of feeling doomed by the mode, he felt motivated by the right tail and determined to fight the disease.

After discussing his bout with cancer, Gould turns to baseball, entering into the great and ongoing debate about the reasons for the disappearance of the .400 batting average; he utilizes the curve of distribution to present a novel theory about this disappearance. In contrast to "good old days" arguments – that baseball just isn't what it used to be – Gould proposes that, when batting averages are approached from the full house perspective, the disappearance of .400 hitting can be explained as a contraction in the range of variation – the distance between the left and right wall – occasioned by the *overall improvement* in play. Notably, as the highest batting averages have declined, the lowest averages have risen; and standardization, Gould points out, often accompanies overall improvement.

It is important to note that the details, and more importantly, the parts of the curve that are emphasized, differ considerably between these first two full houses, disease and baseball (the same will be true for the two full houses yet to come). For baseball, the contraction between walls, a factor that implicates pitching and fielding (as possible components of overall improvement) as well as batting averages, is critical to Gould's point. In the example of cancer survival rates, changes in the distance between the walls are mentioned only indirectly, since the lesson in this case has to do with our reluctance to see the right tail.

While some readers may find either of these first two foci entirely absorbing in themselves, my focus is on the place of these in relation to Gould's final full house: the *totality* of life on Earth considered in evolutionary context – LIFE. In other words I am doing what Gould is urging us to do, i.e., look at the full set, in this case, of his foci. Gould invokes the full house perspective here in an attempt to dislodge the notion of an inherent evolutionary *telos* or drive toward higher forms within the evolutionary process and to dislodge as well the hubris that inheres in such a notion. Although he criticizes, as imprecise and anthropocentric, attempts to define the criteria of the "higher," his main effort is directed toward disproving the notion that evolution is "driven," that is, impelled by an intrinsic force or mechanism slanted toward producing us or other specific complex forms.

Gould opens his discussion of his final full house, LIFE, with a brief discussion of the "drunkard's walk," a textbook example of skewed distribution in a statistical curve. A drunkard is walking down a sidewalk with a wall on the left and a gutter on the right; if he hits the wall he will bounce off and continue walking, but if he strays to the gutter he will fall in and his motion will cease. Despite the fact that the drunkard's motion is random (with no inherent tendency in either direction), the distribution of final outcomes will be right-skewed: the drunkard will always end up in the gutter.

From the drunkard's walk, Gould proceeds quickly to "planktonic forams," a type of single-celled protozoan. He challenges researchers who have assumed that the data on these protozoans indicate an inherent tendency or drive toward larger size through time within this species. Gould counters with the claim that what appears to be a tendency toward increase in size is, rather, a process of random variation in size (in the direction of smaller as much as larger). Analogously to the drunkard's walk, the change *appears* as a tendency toward the larger because the distributional curve is skewed by a left wall, specifically, the minimum size that is necessary for this living form to exist.[2] Those random variations that happen to move in the direction of smaller size quickly run up again this wall and continue no further, while those in the direction of larger size continue onward. Thus, variation in size *appears to be* unidirectional, as if the process of variation incorporates an inherent drive toward the larger. The big lesson common to the drunkard's walk and planktonic forams is that a left wall can impart the *appearance* of directionality to a random process.

Having argued against an inherent drive toward larger size in the case of planktonic forams, Gould goes on to suggest that we reconfigure our overall vision of the totality of LIFE, and the human place in it, on the model of the planktonic forams. Gould asserts a grand analogy: variation within LIFE (that is, across the spectrum of living species) is like variation within the planktonic forams (or for that matter, within the drunkard's walk). For the realm of LIFE as a whole, we should replace concepts of teleological, or "driven," evolutionary variation with the vision of randomly based variation skewed by a left wall of minimum complexity. Larger, structurally more complex forms, such as humans, are merely the right tail of a normal variation around a mode that is bacterial (i.e., bacteria are the most frequently occurring life form), as depicted in Figure 3.2.

Gould aligns his findings with those of researcher Dan McShea, who distinguishes "driven" systems from "passive" systems in terms of the contrasting types of distributional curves each produces (1997:202ff). Gould points out that one can get an indication of whether a system is "driven" or "passive" from the type of distributional curve it generates: in driven systems the mode tends to move rightward and right tails may contract, while passive systems

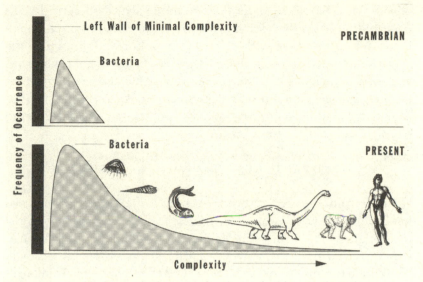

Figure 3.2   Gould's figure. The frequency distribution for life's complexity becomes increasingly right skewed through time, but the bacterial mode never alters.

(i.e., systems based on merely random, rather than intrinsically directional, variation) have stable modes and expanding right tails. Gould's graph of baseball correlates with the driven type of curve; by contrast, the graph of LIFE correlates with the passive type of curve. The driven/passive contrast is pivotal in Gould's work; a thesis implicit in *Full House* is that, in respect to evolutionary process, LIFE is a passive system, or in other words, it is *not* like baseball.

Larger, more complex forms, according to Gould, thus betoken not an upward drive but a random drift away from a mode in the full house of LIFE; we are a "momentary cosmic accident" rather than a "predictable result of an inherently progressive process" (1997:18). Our acceptance of this conclusion, argues Gould, would constitute our biting of the "Fourth Freudian Bullet," an allusion to "Freud's incisive, almost rueful, observation that all major revolutions in the history of science have as their common theme ... the successive dethronement of human arrogance from one pillar after another of our previous cosmic assurance" (1997:17; the passage from Freud can be found in chapter 1).

Freud had emphasized three instances: our cosmic spatial decentring with Copernicus, Darwin's relegating humans to descent from the animal world, and the psychoanalytic discovery of the unconscious. Gould suggests that Freud neglected to consider the contribution of geology and paleontology, the discovery that human life is not co-extensive with planetary time but occupies only "the last micromoment" (1997:18). And besides being late on the scene, any particular right-tailed being is a matter of chance, not a guaranteed occurrence. Gould pronounces paleontology's discovery of our recent and contingent occurrence to be the "temporal counterpart to Copernicus's spatial discoveries" (1997:18).

ADVENTURES WITH MICROCOSMS

Having introduced the framework of *Full House*, I now consider two kinds of problems in Gould's examples. The first has to do with the relation between the different realms that Gould juxtaposes through the curve and his reasons for choosing disease-survival, baseball, planktonic forams (along with the related example of the drunkard's walk), and LIFE in particular. The second problem has to do with the ways in which Gould slips moral valuations into his assessment of the different regions of the curve in his various applications of it.

The Relation between Realms

Gould suggests an analogy between planktonic forams and LIFE, specifically that the processes of variation within each realm are directly analogous. To justify the analogy between his findings regarding planktonic forams and the realm of LIFE, Gould invokes the notion of "self-similarity," a principle that has gained broad attention through the recent wave of interest in chaos theory. "Since we live in a fractal world of self-similarity, where local and limited cases may have the same structure as examples at largest scale, I shall ... argue that this particular case for the smallest of all fossils – single-celled creatures of the oceanic plankton – presents a structure and explanation identical with an appropriate account for the entire history of life" (1997:149).

From a mythologist's perspective, one of the interesting points about self-similarity is that it offers a case of a scientific principle that could easily resonate, in a reader's mind, with a principle that, as attested by countless examples from world mythology, humans with or without science are predisposed to favour. The "folk" cosmologies of the world are permeated with notions of correspondence between the part and the whole, the local and the cosmological: the body (including health, decoration, and disposition); rituals and ritual paraphernalia; houses, storehouses, temples; athletic contests and stadiums; and village or state settlement designs are all typically believed to mirror the shape and rhythms of the cosmos. Now it can easily be objected that these "mythic" notions of self-similarity – positing, as they do, parallels between things of nature and things of human cultural fabrication – are obviously of anthropomorphic inspiration and therefore dismissable as serious candidates for scientific self-similarity. But we face the perils of Scylla and Charybdis here, for dismissal in advance of the possibility of self-similar relationships between human and non-human, or cultural and natural, realms would be tantamount to assuming the existence of a distinctively human substance; and such an assumption would give rise to the charge of Cartesian dualism, which is regarded with at least as much suspicion in the scientific community as is anthropomorphism. The difference in principle between the idea of scientific self-similarity and mythic self-similarity (the latter a species of anthropomorphism) is less obvious than it might initially appear to be.[3] To the extent that Gould's procedure is one of testing whether the relationship between planktonic forams and LIFE is self-similar, it is certainly appropriate that he raise this possibility. But has this relation been proven or assumed? It is also important to keep in mind that the unifying principle that Gould has invoked – replication of patterns at different levels of structure – has a mythologically attested persuasive power and appeal of its own, which antedate by millennia the scientific worldview.

Defining any set, or any particular full house, also raises some interesting issues. Though categories are often fuzzy, each of the first three full houses that Gould considers – survivors of a specific disease, professional baseball batters, and planktonic forams – has the feel of a relatively unified set in which all members are recognizably "variants" of one another. Is LIFE such a set? The term variation is double-edged, for it can assert homogeneity as much as diversity: which of these two connotations will predominate depends upon the context. When invoked in the context of an entity that the reader tends to thinks of as homogeneous, the concept of variation intensifies a sense of diversity. But in realms that the reader less often thinks of in the singular – and for some readers LIFE will be such a realm – speaking of variation within that entity, or of that entity, has the effect of homogenizing and simplifying it, of glossing over differences. Gould, in *Full House* and elsewhere, is fond of criticizing the reifying propensities of biological science; he persistently calls attention to analysts' inclinations to reduce the manifold and contingent to idealized forms that become the "real." But Gould's own move – specifically, his projection of a model of the whole of LIFE drawn from a model of one quantitative parameter (size, which once more recalls Barrow) from one simple species – jumps over the manifold (and some would say "qualitative") differences that are found in LIFE.

All of the realms to which Gould applies the curve in *Full House* – disease survival, baseball, planktonic forams (cum drunkard's walk), and LIFE – are presented as analogous in at least the following minimum sense: they are all sets in which analytical attention to the full house, as opposed to one part of the curve only, will serve to correct previous misunderstandings of the phenomenon in question. The analogy Gould asserts between LIFE and planktonic forams, a correspondence between a vast category and a tiny part of that category, goes beyond this minimum; most notably, the analogy that Gould asserts in this case also includes *process* (i.e., LIFE and planktonic forams both demonstrate passive process; on this issue the analogy also includes the drunkard's walk). But when we move to the relation between LIFE, on one hand, and disease and baseball, on the other, we encounter a more distant kind of analogical relation, in which one of the analogized terms is drawn from everyday, familiar experience, and thus from outside the specifically bio-evolutionary frame that unites LIFE and planktonic forams. And we have already noted that in terms of two general types of curve that Gould distinguishes – driven and passive – baseball and LIFE belong to opposed types. So the question is: does Gould intend any fuller analogies (i.e., beyond the minimum analogy that they are all curve-applicable sets) between LIFE and these more distant realms – disease and baseball – through which he introduces us to the curve and to his notion of a full house?

No, I conclude, but only on second (or third or fourth) reading, since the first reading left me confused on this issue. Even though Gould uses disease and baseball to introduce us to the curve before proceeding to apply the curve to LIFE, the only analogy he ends up affirming between these earlier topics (disease, baseball), on one hand, and the culminating topic, LIFE, on the other, is the abstract, methodological one: that each of these realms can be better understood when analyzed through the curve of distribution. What is common to LIFE and baseball, in other words, is common also to a thousand other phenomena to which a curve of distribution could also be applied. Gould must have other reasons for choosing disease and baseball than that they in particular have something special to teach us about LIFE. Perhaps the disjunction should be unsurprising, since early in the book (although inconspicuously) we are forewarned: "I shall not use the juxtaposition of these examples to present pap and nonsense about how life imitates baseball, or vice versa" (1997:33).

Yet despite this forewarning, a reader could be forgiven for expecting at least slightly more robust analogizing between LIFE and baseball than Gould in the end delivers, for his disclaimer amounts to one terse caveat buried in a lot of talk that could beget other interpretations of his intentions. For example, a good deal of rhetoric in *Full House*, starting with the allusion to the game of poker in the title, might be reasonably taken to suggest that Gould is on the lookout for useful microcosmic metaphors from the realm of sports and games. Especially notable is the opening line of the first chapter: "We reveal ourselves in the metaphors we choose for depicting the cosmos in miniature" (1997:7), which is followed by a discussion of one particular metaphor based on a game – T.H. Huxley's suggestion of chess as metaphor of the natural world – that Gould says is inadequate. The reader might easily assume that Gould intends to offer a replacement metaphor, one based on a different game, for LIFE.

And there is yet one more reason that readers might be inclined to assume that Gould is planning to offer robust analogies between disease, baseball, and the cosmos. It stems from a deep, and likely universal, human mental proclivity. Specifically, the realms that Gould draws upon to introduce us to the curve – the health of the body and the culturally central athletic ritual – are portrayed in mythologies throughout the world as corresponding to the shape, character, and functioning of the cosmos: it is as though they offer a natural launching point for cosmological discourse. By popular acclaim, these are among the world's favourite mythical microcosms, and the source of boundless cosmic imagery.[4] Gould in *Full House* proceeds through these in proper mythological order: self (his recovery from disease), society (baseball), and cosmos (LIFE). What Gould offers his reader is microcosmic progression, content, and rhetoric – accompanied by an inconspicuous denial of microcosmic intent. After introducing some necessary further considerations, I will consider possible motives for such a strategy.

The Coordinates of the Good

At this point I turn to a second, related problem in *Full House*: the shifting moral evaluations that Gould offers in the course of developing his arguments. Gould constructs an elaborate moral rhetoric, focused on the contrast between the mode (or other measure of central tendency) and the right tail of the curve, and resonating with the root problem of the "lower" and "higher" in evolutionary process. I focus on this rhetoric, and particularly a conspicuous mid-book reversal that takes place in the language Gould uses to evaluate the mode and right tail. Specifically, when moving on from disease and baseball to LIFE, Gould inverts the moral valuations of the mode and the right tail. In the two lead-in illustrations drawn from everyday life – disease-survival and baseball – we find a language of praise, desire, hope, and ecstasy directed toward the right tail, and more generally for rightward directionality on the graph. Here we encounter the ray of hope and the cancer patient's refusal to surrender to the mode; here also Gould waxes poetic, if not messianic, about baseball and the .400 average. "But some humans can push themselves, by an alchemy of inborn skill, happy fortuity, and maniacal dedication, to performances that just shouldn't happen – and we revel when such a man reaches farther and actually touches the right wall" (1997:131–2).

Gould, it is true, does argue that we should not lament the rise in the modal quality of baseball performance even though such rises occasion contractions in the range of variation and thus a lowering of peaks such as batting averages, a conclusion that might seem to attest to the value of the mode over the right tail. But in Gould's larger argument about baseball, the right tail still sets the standard: what is good about the rise in modal quality is precisely that the mode, or overall play, has moved in the direction of the peak: the righter, the better. Not only is the rightward movement of the mode a positive thing in itself, but it also creates the possibility of *an even more special right tail*: "But someday, someone will hit 0.400 again – though this time the achievement will be so much more difficult than ever before and therefore so much more worthy of honor … Someday, someone will join Ted Williams and touch the right wall against higher odds than ever before. Every season brings this possibility. Every season features the promise of transcendence" (1997:132). In the case of his disease, the right tail served as a sort of beacon for Gould, galvanizing his will to survive: "I am not a measure of central tendency, either mean or median. I am one single human being with mesothelioma, and I want a best assessment of *my own* chances – for I have personal decisions to make, and my business cannot be dictated by abstract averages" (1997:49).[5] "I saw no reason why I shouldn't be able to reside among these inhabitants of the right tail" (1997:50). Gould's early examples, if generalized for LIFE, would give us the perfect moral scheme for adjudging ourselves and our fellow right-tail (large and complex) beings to be the transcendent part of LIFE. On the model of the .400 batter, we, the few, would be that refined moment that gives meaning to the rest, the tail that wags the dog. However, Gould does not follow this line of thinking.

Moving from baseball to LIFE, Gould instead *reverses the rhetoric*, now taking every opportunity to devalue the right tail vis-à-vis the mode. It is suddenly the mode that contains the true success story; in the case of LIFE this means bacteria, since these are the most commonly occurring form in the distributional curve of LIFE. Consider the tone of these comments on the right tail, in the context of LIFE, as compared with those above, from the context of disease-survival and baseball.

The myopia of characterizing a full distribution by an extreme item at one tail (1997:171 [a sub-section heading])

We are still not free of the parochialism of our scale. If we must characterize a whole by a representative part, we certainly should honor life's constant mode. We live now in the "Age of Bacteria." (1997:176)

the pitifully limited and restricted sense of a few species extending the small right tail of a bell curve (1997:197)

life's evolution, with its massive and persistent bacterial mode, and puny right tail (1997:224)

We cannot confuse a dribble at one end with the richness of an entirety – much as we may cherish this end by virtue of our own peculiar residence. (1997:149)

Such a grandiose claim represents a ludicrous case of the tail wagging the dog, or the invalid elevation of a small and epi-phenomenal consequence into a major and controlling cause. (1997:169) .

There are, then, two rhetorics, one that praises the right tail, seeing it as the epitomizing instance of the whole; and another that praises the mode while speaking dismissively of the right tail. My question is: in any particular application, how does one know which rhetoric to invoke for the curve? Although Gould does not give us an explicit rule on how to choose, the polar rhetorics do correlate with the con-trast between driven and passive systems. And to some extent the two rhetorics of the right tail seem directly motivated, respectively, by the two poles of this distinction; for example, "beautifully orchestrated precision" (1997:130) correlates with striving (i.e., a driven system), while "dribble" suggests passivity (i.e., a passive system). The polar-ity has other dimensions as well – for example, in contrasting assess-ments of the *minisculeness* of the right tail. Specifically, in driven processes, minisculeness is inflected as preciousness ("the pinnacle of excellence is so rare, its productions so exquisite" [1997:130]); here, quality triumphs over quantity, the statistical norm serves only to highlight the exquisite exception. In passive processes, by con-trast, minisculeness is inflected as insignificance ("dribble," "puny right tail"); here, quantity reigns supreme. Seemingly also involved in the move from disease and baseball to LIFE is a reversal of assump-tions about how much the mode is able to tell us about the right tail. In the driven examples of baseball and disease, it is as if the mode, while exerting influence on the right tail, is in the end unable to account for the latter; the fact that the tail *stands out from* the central tendency is what is important – an orientation epitomized in Gould's claims in regard to his disease: "I am not a measure of cen-tral tendency" (1997:49). We do not see Gould making an equivalent proclamation – "We are not bacteria!" – regarding our place in LIFE; indeed the emphasis in the case of LIFE is on how well the mode *does* account for us, since the process that explains the tail (which includes humans) is nothing more, according to Gould, than an extension of the process that explains variation within the mode itself.

Finally, toward the end of the book Gould elaborates on how we "yearn" to assign different meanings to the same consequences according to whether they are *"results directly caused"* or *"consequences incidentally arising"* (1997:196), once more suggesting that the reversal of rhetoric hinges on the driven vs. passive contrast. For these reasons, and because I find no other, I conclude that it is the driven vs. passive distinction that, for Gould, authorizes the reversal of rhetoric when he moves from baseball to life. I will refer to this reversal of the rhetoric of the right tail and mode as Gould's "grand flip" – borrowing a phrase coined by Gould (1997:41) to dramatize the radical change he attributes to Darwin regarding the way we conceptualize variation, and ultimately progress or its absence, in nature. But if Darwin's flip was a matter of change, Gould's flip is a mechanism of continuity: specifically, Gould maintains his self-consistency on the idea of progress in evolution (a possibility he rejects) by spectacularly flipping the evaluative rhetoric he brings to the mathematical instrument through which he portrays evolution, i.e., the curve of distribution, and specifically its mode and right tail.

PROBLEMS WITH GOULD'S "GRAND FLIP"

Gould's grand flip is full of quirks, most of which arise from his attempt to inject into the exposition of science his enthusiasm for life; manifestations range from boyish Gee Whiz! to heroic adventure and high art. All of these quirks work to undermine Gould's claims, coming close, in some cases, to toppling them. The following are the more obvious quirks.

1. There is an implicit reasonableness in heaping laudatory terminology on the right tail (or the rightward direction in general) in many driven systems, just because the right represents the region toward which the system as a whole is striving. We tend to think in contrasts, however, and thus may be tempted to conclude that the reverse will hold true for the right tails of passive systems: dismissive language, the reverse of laudatory language, should apply to these. But such a conclusion is a logical non sequitur, akin to deriving "all non-men are non-mortal" from "all men are mortal."

It would be technically incorrect, of course, to apply to passive systems *that part* of laudatory terminology that implies willful striving within the system. But in our ordinary language and experience, a great deal of laudatory terminology – terminology of the "rare" and "exquisite" – implies no position on the driven/passive issue: things achieved by human willful striving, whether in athletics or opera, can be rare and exquisite; so can objects or events resulting from random forces, such as glittering gems, spectacular sunsets, the subtle patterning of water in motion, and moments of good fortune that may "come our way" without our seeking them.[6] Should we add savannah-like environments? The movement of evolutionary aesthetics – of which we saw glimmers in the previous chapter – attempts to locate the source of aesthetic response in Darwinian selection: in striving to produce art, humans are in effect attempting to recreate, through will, what has already been created by differential survival in the context of evolution. Within evolution there may be processes that are driven in Gould's sense, but let us not forget that Gould is trying to convince us that processes that we might prefer to see as driven, especially those connected with our own origins, can turn out to be passive. In sum, the rules, in ordinary language, of whether to invoke laudatory vs. dismissive discourse do not derive solely from the driven vs. passive contrast. What are the rules on the cosmic level? One response, of course, could be that the laudatory language of baseball and opera, and even that of diamonds and spectacular sunsets, belongs only to the human world and has no place in cosmic commentary. However, the analysis that follows will reveal that Gould's dismissive language is no less saturated with the experience of the everyday human world. Hence, by the same reasoning, Gould's dismissive language, too, should have no place in the cosmic commentary.

2. Following from the previous consideration, it is important to note that Gould seems to see human cosmic hubris as bound up almost exclusively in the belief some people hold that evolution is *driven* toward us, that we are an inevitable and necessary consequence of evolution. At one point Gould portrays an imaginary reader reacting to the idea that we came about through processes that are radically contingent; this reader attempts to retrench around a lesser form of the claim that evolutionary process is a process driven toward us: "The right tail may be small and epiphenomenal, but I love the

right tail because I dwell at its end ... Even you admit that the right
tail had to arise, so long as life expanded. So the right tail had to
develop and grow – and had to produce, at its apogee, something
like me. I therefore remain the modern equivalent of the apple of
God's eye: the predictably most complex creature that ever lived"
(1997:174). To the reasoning of the imaginary opponent, Gould
responds: "Wrong again, even for this pitifully restricted claim (after
advancing an initial argument for intrinsic directionality in the basic
causal thrust of all evolution). The right tail had to exist, but the
actual composition of creatures on the tail is utterly unpredictable,
partly random, and entirely contingent – not at all foreordained by
the mechanisms of evolution" (1997:174–5).

Now, it is no doubt true that for some individuals who believe that
humans hold a special place in the cosmos, this belief is bound up in
the view that we are in some way a necessary and predictable outcome of evolution. For others, however, the problem of the human
place in the cosmos, and any sense of a special place as well as any
hubris that may flow from it, has less to do with our alleged necessity than with the capacities that some species possess and others
do not – and, as regards humans, pre-eminently such capacities as
self-consciousness, language, and analytical reason. Gould likes to
analyze the ways in which human yearnings skew interpretations
of biological evolution, but his analyses of human yearnings are
themselves skewed in ways that support his case. Consider the following example offered by Gould: "The purposeful killer and the
erring policeman produce the same result ... yet we yearn to judge
the meaning differently based on the distinction between intention
and accident" (1997:196). Gould moves from this example directly to the contrast between driven vs. passive process in evolution:
"Our intuitions detect a radical difference in meaning between these
two pathways to predictable production of the same result – and
our intuitions are right again. We do, and should, care profoundly
about the different meanings – for, in one case, increasing complexity is the driving *raison d'être* of life's history; while, in the other, the
expanding right tail is a passive consequence of evolutionary principles with radically different main results" (1997:197). The shooting scenario offers, in effect, yet another microcosmic analogy – one
that, like the planktonic forams and drunkard's walk, Gould thinks
will be helpful: as we assign a different meaning to a killing by a
purposeful killer and an erring policeman, so we should assign a

different meaning to the idea of driven vs. passive evolutionary process. Gould would like us to accept this analogy so that he can use it to deflate us.

But just what meanings do we assign to the two killings? It is true that people may feel differently about the same bad consequence depending on whether it was intentional or accidental, but not necessarily in a uniform way. In informal polling on Gould's scenario, I discovered that some people found that accidentalism made the consequence of shooting death in Gould's scenario *less bad* (because of lack of evil intent), but others thought it made the consequence morally *worse* (because it seemed even more senseless). More importantly, whatever the differences, no one proposed that the intention vs. accident distinction was enough to change our assessment of the fundamental character of the consequence itself: the difference did not turn a bad consequence into a good one. In our legal system, the intention vs. accident distinction can alter fundamentally the character of a harmful act, changing it from bad to neutral, but this is a judgment about the agency, not about the consequence – the latter remains fundamentally bad. By analogy, then, the moral character of our place in the cosmos would be fundamentally what it is irrespective of the agency (or lack of it) through which we came to be. Just as knowledge of the accidental character of a shooting death does not erase the misfortune, so, one could argue, knowledge of the accidental character of our evolution does not erase the good fortune.

At the same time, Gould's linking of the problem of cosmic hubris with the issue of drivenness vs. passivity ends up abetting other possible forms/sources of human cosmic hubris. Just as accidentalism in the shooting scenario for some people made the bad even worse, so accidentalism can make the good even better. Hammering away against the idea of our necessity can actually feed the hubris rooted in convictions about our possession of rarer capacities, because radical contingency makes them just that much rarer. One commentator has noted that Gould is popular among some creationists, who "for their own philosophical reasons, want to depict the evolution of a human level of intelligence as spectacularly unlikely" (Wright 1999:59).[7] And of course the more contingent the processes, the more unique the outcomes; and uniqueness itself tends to be positively valued in our individualistically inclined society.

In sum, the ways in which and the extent to which human hubris is bound up with the driven vs. passive distinction are extremely complex and highly debatable; Gould spins these matters in ways that are selective and maximally support his larger agenda.

3. In considering Gould's portrayal of the curve of LIFE, it is important to lay bare the criteria that result in the mode coming off better than the right tail. The criteria come down to one factor: quantity. As noted earlier, if one compares the examples, cited above, of laudatory rhetoric for the right tail (in the case of baseball) vs. dismissive rhetoric for the right tail (for LIFE), one sees that the laudatory rhetoric is largely framed in qualitative terms while the dismissive rhetoric is largely in quantitative.

Although Gould says more than once that he does not want to fall into the "silly mythology of 'bigger is better'" (1997:90, 158), the idea that *more is better* apparently is not, in his view, silly mythology, for that is what Gould gives us: the bacteriological success story as measured by quantity – numbers, ubiquity, and total biomass. "Not only does the earth contain more bacterial organisms than all others combined (scarcely surprising, given their minimal size and mass); not only do bacteria live in more places and work in a greater variety of metabolic ways; not only did bacteria alone constitute the first half of life's history, with no slackening in diversity thereafter; but also, and most surprisingly, total bacterial biomass (even at such minimal weight per cell) may exceed all the rest of life combined ... Need any more be said in making a case for the modal bacter as life's constant centre of maximal influence and importance?" (1997:194). Gould delights in the remarkable properties of bacteria, but the remarkableness in each case comes down to adaptive mechanisms that allow bacteria to be "dominant" – and dominant seems to mean that there are a lot of them (see all of his chapter 14, especially 169–98).

But if basic weights and measures are the most obvious indication of the dominant role played by quantity in Gould's assessment, other manifestations are more intriguing. Gould uses a number of terms and analytical devices that make light of structural and other qualitative distinctions between different living things. For example, besides lack of importance and passivity, "dribble," a term Gould invokes to put down the right tail in the context of LIFE, tends to imply a seepage of one continuous substance as opposed to a process of transformation; in Gould's usage, dribble neglects the fact that entities occurring in the right tail of LIFE are structurally distinct from those of the bacterial reservoir from which they seemingly drip.

At one point Gould's quantitative considerations bump up against something like qualitative ones. Specifically, regarding the tendency of humans to place themselves at the pinnacle of evolution, Gould speculates that, "any truly dominant bacterium would laugh with scorn at this apotheosis for such a small tail so far from the modal centre of life's main weight and continuity. I do realize that bacteria can't laugh (or cogitate) – and that philosophical claims for our greater importance can be based on the consequences of this difference between them and us. But do remember that we can't live on basalt and water six miles under the earth's surface, form the core of novel ecosystems based on the earth's interior heat rather than solar energy, or serve as a possible model for cosmic life in most solar systems" (1997:198).

By abruptly bringing to a close his rumination about scornful bacteria following the observation that bacteria can't laugh or cogitate, Gould stops just short of giving us a *fable* – that of the bacterium and the human – to set next to other fables about improbable overturnings of hierarchies, for example, the lion and the mouse, or the tortoise and the hare.[8] From Gould's account we might conclude that the reason for his terminating the fable was an attack of realism. But it should also be noted that had Gould followed his fable out, it would have given rise to dilemmas and ultimately backfired on him. Traditional fables of improbable overturnings invariably pit brawn vs. brains, i.e., beings of greater physical size or prowess against beings with less imposing physical credentials but greater mental agility. The former are scornful of the latter initially, and then get their comeuppance as the physically inferior being has the last laugh. A bacterium is smaller than a human, and since bacteria receive the praise while human pretensions are put down, the outcome of Gould's developing fable is consonant with the traditional fable. Moreover, bacteria, in Gould's portrayal, have the last laugh – until Gould admits that bacteria can't laugh. The plot unravels further, for in terms of the way that Gould in this book thinks about mass – relative bio-dominance as judged by comparison of the total mass of different species – Gould's *bio-dominant but non-cogitating* bacteria would make the better lion, and we, the "puny," non-bio-dominant but cogitating right tail, would make the better mouse. That Gould awards the contest to the (collectively) large but unthinking species, bacteria, neatly summarizes his criterion of assessment, which in the end reverses the traditional folk wisdom of the fable: in judging evolutionary success, Gould goes by brawn rather than brains. Surely this criterion of success contributed to the abrupt termination of the fable-in-the-making!

Indeed, the mention of laughing bacteria opens the problem of possible qualitative distinctions in the curve of LIFE, for which Gould's analytical perspective has nothing to offer. Gould's response to the problem he raises with his proto-fable consists solely of a list of capacities that allow bacteria to be numerous (perhaps even having a cosmic distribution). Adopting the lion's point of view, in other words, Gould quickly buries the possibility of qualitative distinctions under biomass. A similar pattern occurs in Gould's chapter 12, in which once again a debate on the problem of defining the "higher" in evolutionary process is terminated abruptly with an observation about the ability of ants and bacteria to thin our numbers: "If progress is so damned obvious, how shall this elusive notion be defined when ants wreck our picnics and bacteria take our lives?" (1997:145).

There is a third, though somewhat different, instance of the same pattern; specifically the last sentence of the book (excluding the epilogue on human culture). The sentence begins, "we are glorious accidents" (1997:216). But whence this "glorious"? Why a strongly laudatory, qualitative term applied to a being of the mere dribble? How can this be reconciled with the earlier dismissive rhetoric? I see two possibilities. One is that the "glorious" points us to the epilog, which is about human culture and driven excellence within it: under this interpretation the comment is not inconsistent because it is not about the curve of LIFE but only the realm of human-specific, driven excellence such as we find in baseball. In the other interpretation, the comment follows the pattern of the two previous examples (the non-cogitating bacteria and the discourse on the higher that turns into the problem of ants and picnics). That is, the "glorious" forms the culminating instance of Gould raising the spectre of qualitative differentiations within LIFE only to abruptly terminate the conversation.

Gould does accept that passive right tails will be characterized by the appearance of organisms of ever-greater complexity, but he argues that greater complexity in itself is not grounds for seeing "progress" in evolution. The occurrence of organisms of ever-greater complexity would be progress only if the greater complexity could be shown to result consistently in the increasing dominance of more complex species; in that case, we could conclude that complexity bequeaths superior adaptation, that complexity is therefore selected for, and that evolution is therefore driven toward "inherently advantageous complexity" (1997:173). (I add that if we were able to conclude that evolution is driven toward complexity, we would also then have, according to Gould's criteria, legitimate reason to think of ourselves as rare and precious; and, as we do in baseball, we would start fretting about the *appearance* of degeneration in our species occasioned by the improving mode of LIFE.) But, following Gould, we cannot conclude any of this, because we do not have the numbers to prove that greater complexity produces greater biodominance. We must, therefore, stay with the dismissive rhetoric of the right tail, foregoing baseball talk in the context of biological evolution. Gould's arguments reflect a moral vision common in the post-Darwinian world, in which our cherished human values fall as species-centric fancies before the one value objectively given in nature: reproductive success, which is quantifiable. We tend to think of numbers as objective, and the fact that reproductive success can be quantified may lead to the unfounded conclusion that the value itself – the state or achievement that is valued – is an objective value.

Gould's quantitative bent – evident in the several ways discussed above – finds a fitting summation in the spectacle of the great curve of LIFE. This diagram stands for the whole while advancing certain propositions about the relationship of parts within it; and as a quantitative device the great curve conjures up the authority of science. Indexing a form of technical mathematical analysis foreign to traditional mythologies, the curve nevertheless functions like many symbols in myth, offering, like the mythic world parents of Sky and Earth, an emblem of the unity of all life. But although Gould's cosmically integrative emblem points to techniques generally acknowledged to distinguish science from myth – specifically, technical mathematical/quantitative techniques – it is important to recognize that, considering the immense reach of the curve of LIFE, Gould's "quantitative" analysis is rather short on number crunching. Inspired by a quantitatively analyzed microcosm (the planktonic forams), the great curve of LIFE is, for the moment at least, still in large measure a speculative cosmic projection of that microcosm – with vast stretches of unquantified LIFE brought under its aegis through a variation on the "quantitative impressionism" discussed in the previous chapter.

4. Since Gould's aim is broadly to lead us out of visions of the cosmos distorted by our human desires and aspirations, it is appropriate to ask whether Gould's own formulations are free of those inclinations. The influence of political and cultural views on scientific formulations is one of Gould's own favourite themes; it appears in *Full House* in an analysis of the ways Darwin's political views sometimes led the master astray (inducing the earlier mentioned grand flip) on the issue of evolutionary progress, and it surfaces at numerous other points in his writings, notably in *Wonderful Life* (1989),[9] which Gould calls the companion volume to *Full House* (1997:18).

In the epilogue on "culture" in *Full House*, Gould notes that as a "tough-minded intellectual" (1997:229) he hopes that *Full House* will not be turned into a "shill" for the doctrine of political correctness. The cultural and political resonances that show through in *Full House*, however, are less those of political correctness than those of political populism. The rhetoric that Gould draws on to praise the tail (and improving mode) in baseball and opera is of course a rhetoric already familiar to the reader before reading *Full House*. But, I insist, so is the rhetoric that praises the mode of LIFE – bacteria – while dismissing the special claims of the tail. Gould's bacteriocentrism, like Xenophanes' hippocentrism, if not anthropocentric, is at least anthropomorphic, emerging originally not in bacterial cultures but human ones. Specifically, the rhetoric of Gould's bacteriocentric moralizing mirrors the rhetoric that has been used historically to extol "ordinary people," the "common man," the "folk," the "lower classes," the "good people of America." The latter possess the collective virtues of Gould's bacterial mode: they form the stable, resilient, and persistent centre of life; they carry on a vigorous, variegated existence beneath the gaze of the "sophisticated," and some of them even live underground. They are the source of a society's continuity and basic character. That Gould's anthropomorphized bacteria are so impressed by quantity also might tell us something about the kind of people bacteria are. The two different value systems that Gould has invoked – one to tell us why .400 batters and Pavarottis are great; the other to tell us why bacteria are great – are the same polar value schemes that we use to glorify, on one hand, human elites and, on the other, those humans that we designate as the *"masses"* – a telling designation.

One sometimes encounters the human version of this polarity expressed in spatial terms. Modal values are particularly associated with the geographic centre – the (anthropomorphically inflected)

"Heartland" – and right tail values with the coasts (originally a West-Coaster myself, I do not share the view that the West Coast is a left wall or minimum criterion of culture). In discussing the curve (1997:55), Gould mentions a micro-level spatial distribution, specifically the potential cultural bias in the long association of "right" with "higher" in much of Western tradition. To those biases Gould mentions, we can add the political connotations, which fit the graph rather nicely (e.g., the left's solidarity with the masses; the right's association with economic elites).[10] One can, of course, make too much of such directional associations. They are interesting, and betoken a strategy humans inevitably resort to in representing contrasting values; yet such spatial coordinates are a less important consideration here than are the values themselves.

The polar value schemes Gould invokes for right tail and mode – and which he reverses when moving from baseball to LIFE – would seem to be, rhetorically at least, one particular dispensing of the dual strains of embracive populism and cosmopolitan elitism that run through all of Gould's work; his writings reveal a passionate commitment to democratizing science combined with a fondness for letting the reader know which object of European high art has triggered his latest musing.[11] But Gould need not be singled out: some version of the complicated combination is clearly the statistical modal moral/political position among American university academics, if not among humans more generally. Where the two standards coexist they solidify one another even in their opposition.

And if the two perspectives themselves are familiar, the strategy or process of reversing them – shifting from one perspective to the other – is also an experience familiar to most if not all humans. A good deal of anthropological and sociological literature suggests that reversibility is intrinsic to virtually all social hierarchy. At one moment the pinnacle of the hierarchy is stressed; the next moment, in a "ritual of reversal," the pinnacle is downplayed and the masses celebrated, the ruler becoming nothing without the masses, who in their collectivity are the true source of power and social identity. Hierarchy reversals can be inflected loftily ("blessed are the meek, for they shall inherit the earth") or with the carnivalesque appeal of the wrestling ring or the court jester.

I emphasize the process of hierarchical reversal as a necessary corrective to Gould's own – unfortunately inaccurate – elaboration on what he means by his phrase "full house." The full house perspective, as depicted in the abstract by Gould, involves learning to see and consider the entire range of variation: "we must abandon a habit of thought as old as Plato and recognize the central fallacy in our tendency to depict populations either as average values (usually conceived as 'typical' and therefore representing the abstract essence or type of the system) or as extreme examples (singled out for special worthiness, like 0.400 hitting or human complexity)" (1997:3–4). In abstract, generalizing moments such as this, Gould thus distinctly conveys the impression that he seeks some sort of synoptic gaze that takes in "the richness of an entirety" (1997:149) and escapes the inclination to single out any particular segment of the curve for special attention. Gould mentions two parts of the curve that we tend to single out for special attention: one is the "average" value (which elsewhere he calls "central tendency" or shortens to one particular measure of central tendency, i.e., "mode"); the other is the extreme example (which elsewhere he calls the "right tail"). It is important to note that for Gould a focus on *either* of these segments, rather than on the full curve, amounts to Platonizing; the opposite to Platonizing would be to let go of a special focus on *either* of these segments in order, instead, to take in the whole.

However, when Gould turns from abstract theorizing to analysis of specific full houses, he tries to convince the reader of something quite different. Gould *continues* to focus on the same two parts of the curve – mode and right tail – that, he claims, we have wrongly singled out all along. He does not even try to shift the reader's gaze to "the richness of an entirety"; instead, he tries to convince readers to *reverse* the hierarchical assumptions that we brought (or that Gould assumes we have brought) to the particular full house under consideration. For each full house, Gould tries to convince the readers that the part (whether mode or right tail) of the curve that we traditionally thought was *least* worthy of our attention is really the *most* worthy. The "entirety" is never addressed, and Gould has nothing at all to say about the poor devils who dwell in the hinterland, between mode and right tail.

To be specific, the case of disease opens with Gould supposing that the usual reaction to cancer survival statistics permits the mode to dominate: the average victim of this disease will be inclined to assume that his/her survival will fall somewhere in the mode. Pointing out that measures of central tendencies are "abstract measures applicable to no single person, and often largely irrelevant to individual cases" (1997:48), Gould shifts the focus to his determination to "reside among these inhabitants of the right tail" and relates how this resolution shaped his life (1997:50).[12] In other words, focus on the mode is replaced by focus on the right tail. The case of baseball involves a double-reverse. Gould opens by reviewing degenerationist explanations of falling batting averages and argues that dispirited sentiments arise because of our fixation on the right tail. To counter such sentiments, Gould points out that as the highest averages have fallen, the lowest have risen; this and other factors suggest that the process of the right tail shrinking is more accurately described as a process of modal play improving. But after arguing that we should all celebrate the improving mode, Gould once again returns to and concludes his analysis with the messianic passage quoted earlier, depicting the eventual return of the – now even more special – .400 average. So, in other words, we go from focus on the right tail, to focus on the mode, and then back to the right tail. Gould opens his case about our place in LIFE on the assumption that most readers will not as yet have bitten the fourth Freudian bullet, or in other words, that they will retain at least a lingering sense that evolution enshrines us in the highest position; against this view, Gould awards that place to the bacteria (more on this below). In other words, in this instance we shift from right tail to mode. In each of these three cases – disease, baseball, LIFE – we end not with the "richness of an entirety" but merely with a reversed (or double-reversed) asymmetry of focus. We have in no way gotten beyond the problem of Platonizing as Gould has defined it. Grasping the "richness of an entirety" (as opposed to merely reversing hierarchies) is apparently as difficult to achieve cosmologically as it is socially.

The fact that *in practice* Gould's full house means not equitable synoptic gaze but rather hierarchy reversal, requires us to ask a further question: does Gould really lead us out of anthropocentrism? Or does he give us something more like *reverse anthropocentrism* – a sort of bacteriocentric anthropo-bashing – that, whatever its honourable goals, ends up projecting onto the biosphere the same dilemmas and potential for miscarriage of intention that are reflected in North American discourse on social equality, specifically in debates about whether policies announced as promoting equality might in some cases be policies of reverse inequality? The root problem is that even very sophisticated thinking about hierarchy reversal does not necessarily make a positive contribution on how we are to envision "the

richness of an entirety" (surely Gould is trying to envision something more than a system of *rotating pre-eminence* between mode
and right tail, which, again, would have nothing to say about those
that lie *in-between*, on the periphery of both). Is it perhaps part of
the popular slant of science writing to follow familiar and ultimately
anthropomorphic tactics in attempting to redress prejudice? Humans
are intent on inscribing cultural and political values on the cosmos,
seemingly, at least in part, because of the moral and rhetorical power
of doing so: human, all too human, values and moral strategies then
can be seen as a derivative, a microcosm, of principles that have cosmic backing and force.[13] Such cosmic visions are about *us*. We are
rightly warned by Gould and others to be concerned about the possible influence of anthropocentric bias when we encounter visions
of the cosmos or of evolutionary process that place us humans at
the pinnacle. But this does not mean that visions that do not put
us at the pinnacle are ipso facto free of anthropocentric motive or
anthropomorphic inspiration. There is no reason to assume that the
inclination to inscribe human values onto the cosmos or onto the
structure of the biosphere applies to elitist values only.

5. The insistent focus on the contingency of the right-tailed deni-
zens has another important consequence in Gould's analysis, this
one rather surprising: having opened *Full House* by attributing our
wrong-headedness about evolution to lingering Platonism, Gould's
book culminates in a vision of LIFE that has a distinctly Platonic
flavour. Historians of science often portray Darwin as delivering –
in the idea of "transmutation" of species – the death blow to the
Platonic conceptualization of species, and Gould's views align with
Darwin's in this respect. Yet what Gould takes away from life at the
level of species, he gives back at the level of LIFE.

Gould claims that Platonism lingers on in our predisposition to
see only part of the curve of variation rather than the full house;
and as already noted, such lingering Platonism, according to Gould,
can take the form of our being mesmerized by *either* the mode or
the right tail. Yet in the midst of his analysis of LIFE, Gould says, "if
we must characterize a whole by a representative part, we certainly
should honor life's constant mode" (1997:176). This comment is followed by a bacterial encomium that goes on for some twenty pages.

But the very choosing of a representative to stand for the whole of
LIFE manifests the lingering Platonism that the full house perspective

supposedly opposes. As if aware of the contradiction, Gould announces his choice in a reluctant ("if we must") tone; so perhaps despite its scope and enthusiasm the encomium is not intended all that seriously. On the other hand, however, it will be recalled that in every previous example, Gould's insistence that we consider the full house amounted, in practice, to the same move that Gould is making here: shifting our focus from one part of the curve to another (in this case, from right tail to bacterial mode). Whatever Gould's broader intentions, it is also important to note that, in deciding which part of the curve should represent LIFE, Gould's argument – judged by his own announced goals – is exactly backwards. Specifically, in regard to the goal of getting rid of lingering Platonism, Gould, in making the reluctant choice, would have done better to choose humans or another denizen of the right tail, rather than the bacterial mode, to serve as the representative of LIFE. Why? Because the pre-eminent quality that Plato sought in his doctrine of "Forms" was constancy or indestructibility. The exemplary – the *true* – for Plato inhered in imperviousness to change, particularly degenerative change. If we really want to shed our Platonic bias, we could do no better than to single out for honours one of the freak species that burn brightly for a cosmic instant and then are gone (and are unlikely to have ever been at all). Following Gould's own characterization, that could be us! Indeed, such wild contingency is the characteristic of life that Gould chooses to celebrate in his companion book, *Wonderful Life*.

In choosing bacteria for the honours, not only does Gould choose the more Platonic representative to stand for LIFE, but he conjures up a Platonic aura around his choice, going so far as to invoke (and tweak) a venerable liturgical formula, to which the Platonic background of Western theology may not be irrelevant: "Life still maintains a bacterial mode in the same position. So it was in the beginning, is now, and ever shall be – at least until the sun explodes and dooms the planet" (1997:170).

The deep contrast that impels Plato – between something like eternal principle and accidental excrescence – also haunts Gould's *Full House*. Plato's *Republic* and Gould's *Full House* are equally permeated by the theme of degeneration and its remedies. The so-called "guardians" of Plato's *Republic* – protected from the unseemly myths told by nursemaids, and schooled instead in the pursuit of the eternal Forms through philosophical dialectic – would steer the body politic clear of the degenerative impulses that beset civilizations and engender myths. In kindred spirit, virtually every section of *Full House* involves the theme of degeneration, whether in the form of disease, baseball, or the waxing and waning of bio-evolution-ary lines as suggested by "lingering vestiges of former robustness" (1997:64). In each case Gould utilizes the curve of distribution to propose a remedy to degeneration: in the case of disease, the ray of hope in the right tail; in the case of baseball, the realization that declining extremes can actually reflect overall improvement; in the case of LIFE, the shifting of our gaze to bacteria as the true source of life's constancy and the recasting of bacteria, in place of ourselves, as the venerated representative of LIFE. True, Plato, convinced of the corruptibility of all corporeal matter, spoke of the Forms in other-worldly terms. Lacking the findings of modern microbiology, he was unaware of the principle of constancy to be found just beneath his feet (and all over them). Had Plato been a modern biologist, he would have been Gould.

In *Full House*, Gould wants to undo distortions of the cosmos that are introduced by "*our yearning for general progress – that is, the predictable and sensible evolution to domination of a creature like us, endowed with consciousness*" (1997:174). But, in addition to this particular yearning, humans have other yearnings whose effects in shaping our views of the natural world, and indeed our sciences, we should consider. Among these is a cross-culturally attested human yearning for models of constancy in the cosmos, a desire expressed not just in Platonic philosophy but also in countless traditional myths and rituals, many of which are based on the theme of eternal return to a primordial cosmic paradigm. Just this infatuation with cosmic constancy has often been cited as the source of the human inclina-tion to mythic-mindedness. One recurrent theme among eighteenth- and nineteenth-century mythologists, for example, is the claim that over-regard for constancy, especially in the form of reverence for "tradition," is the great force responsible for humanity's being mired in myth rather than progressing toward scientific enlightenment.

The concept of "myth" winds its way through both Plato's and Gould's speculations about why we fall into error and how we find our way out. The tradition of invoking the term "myth" to exemplify the defective point of view, the view distorted by inflated human self-regard, is by now of venerable antiquity in Western intellec-tual history. And Plato more than any other thinker is responsible for this view of myth – the view that Gould in *Full House* invokes in polemical phrases such as "such mythology is not an option for thinking people" (1997:19; cf. 80ff).[14]

Both Plato and Gould seem to regard story as a cognitive vehicle particularly susceptible to anthropocentric inclinations. And, although the specifics are admittedly very different, there are intriguing parallels in the visions that Plato and Gould offer, respectively, to correct the self-infatuation of the mythological view. The self-infatuation that concerned Plato (and Xenophanes) took the form of a readiness to portray the gods in human image (with Homeric gods as exemplar); to this Plato counterposed the idea of an eternal, stable deity that was rational rather than fickle. For Gould our self-infatuation takes the form of a readiness to portray ourselves at the apex of nature. But both Plato and Gould seek to remedy our self-infatuation by directing our attention to forms of cosmic constancy superior to our own. In terms of "dominance" – the only value criterion that Gould allows into his discussion of "progress" – bacteria are the ideal form. We, on the other hand, are part of the "pool of accidents" (1997:41) – a phrase that recalls the classic philosophical distinction of "essence" and "accident" and that Gould uses to characterize Plato's notion of degraded imitations of Forms; the phrase also captures rather well Gould's own view of the right tail of LIFE, nicely continuing the dribble metaphor.

A final and particularly intriguing parallel lies in the fact that, identifying myth with the defective view, neither Plato nor Gould entirely foreswears mythical means of persuasion. Plato accords legitimacy to the myth or "noble lie" that holds society together,[15] and he constantly invokes mythic imagery in explicating his own teachings – notoriously, the imagery of passage from inner to outer, darkness to light, in his famous "Allegory of the Cave."[16] And as I suggest throughout this chapter, many aspects of Gould's presentation – the heroizing, the storytelling, the playing with microcosm/macrocosm relationships – are familiar turf to the mythologist.

6. In *Full House*, Gould aims to give us an objective account of our place in the cosmos, the objectivity of which turns on quantitative considerations. But seemingly as important as the objective account itself is the moral lesson Gould attaches to it: we will be dethroned from the central place that in our arrogant yearnings we have constructed as mythological worlds. Why it is important to Gould that our arrogance be refuted is not entirely clear; if it is

solely in the interests of objectivity – that we will be more objective if we are less arrogant – then the following comment is amiss and can be disregarded.

In contemporary culture, appeals to consider the other person's point of view are typically calls to reassess one's own moral vision. For this reason among others, I get the feeling that Gould, in portraying evolution from a bacterium's point of view, is advancing a moral claim – specifically, that we will be not only more objective people but also morally better people if we give up the idea that evolution is propelled by a mechanism of progress and that we are evolutionarily more highly progressed than bacteria. If Gould is making this claim, then I am not convinced.

A demonstration that greater complexity generally results in expanding biodominance (thus indicating an evolutionary mechanism that selects for complexity) is the only proof that Gould says he will accept of the proposition that structurally more complex beings are more highly progressed beings. Put another way, Gould is saying that evolutionarily we are not more highly progressed than bacteria, but that if we were, the validation would lie in our (and other complex species') consistently expanding dominance of the biosphere (which, by his own logic, would include killing other species and wrecking their picnics to deflate *their* claims to being more highly progressed). But *morally* speaking, does the demonstration that we are not more highly progressed humble us when the criterion of progress is itself amoral? Part of what Gould seems to offer to us with the curve of LIFE is a morality lesson. However, in the end we have to ask: is Gould's new, humbling vision of our place in nature (which would credit us with progress only if we increased our biodominance and spoiled other species' picnics) more morally compelling than the older, supposedly arrogant vision – the one that, rooted in the doctrine of progress, gives us credit, points even, for our ability to think?

THE RATIONALE FOR FAUX MICROCOSMS

Let us now return to the issue raised earlier, the pattern in which Gould progresses from the local to the cosmic: the health of the body, the culturally central athletic contest, and finally LIFE (with the drunkard's walk and planktonic forams slipped in tersely as a preamble to LIFE). If Gould invokes the curve of distribution specifically in order to offer us a view of the evolution of LIFE as passive rather than driven, then why does he lead us to this conclusion through realms of experience – fighting disease, baseball – that are driven? Let me phrase my question in another way, and fill it out a bit more. If we list the *familiar everyday* illustrations used by Gould to draw us into his vision of the cosmos, we find (excluding asides) four: disease survival, baseball, the drunkard's walk, and the game of poker, to which the book's title alludes. Of these four, three – disease survival, baseball, and poker – have at least partially driven right tails; indeed these are all contests or struggles oriented toward a pre-existent goal, and thus diverge from the alleged "random" right tail of LIFE. Only one illustration, the drunkard's walk, mirrors LIFE's random right tail. Yet the three driven examples all appear earlier and more prominently – and are more fully savoured – in *Full House* than is the drunkard's walk. Disease survival is "Case One" of the curve; baseball furnishes the subject matter for the most elaborate section prior to LIFE; poker furnishes the book's title and dominant metaphor and is also the medium through which Gould, in the mock just-me-and-you tone characteristic of mass-oriented self-help writing, opens his book by offering the reader a wager (one, incidentally, with considerably more risk for the reader than for the writer): "So let me make a deal with you. As a man who has spent many enlightening, if unenriching, hours playing poker (hence the book's title), I want to propose a bet. Persist through to the end, and I wager that you will be rewarded (perhaps even with a royal flush to beat my full house)" (1997:2). By contrast, the drunkard's walk is slipped in late in the game and dispensed with rather quickly, even though it is the most analogous example, on the critical point of drivenness/passivity, to the evolutionary vision of LIFE offered by *Full House*. So why doesn't Gould open with and linger on the drunkard's walk? And, of course, why doesn't Gould entitle his book *The Drunkard's Walk* (which contains, after all, the book's ultimate message)? Doing these things would obviate the need for the grand flip that introduces the rhetorical complications discussed above.

There may be more than one reason why the grand flip prevails in Gould's book. Most obviously, the juxtaposition of baseball and LIFE may help the reader to understand the inclination to transfer the interpretation of one system to the other – as background to Gould's implicit arguments about why we should not do this. If this is Gould' intention, his coyness about it begets confusion. Moreover, there is much more baseball in *Full House* than is necessary for making such a comparison, and so I want to suggest an additional motive. Specifically, I suggest that the flip, on the matter of drivenness, between homey opening examples and evolutionary macroperspective, is necessitated by the very teleology that drives this book. Gould is trying to interest readers in an unfamiliar perspective, and he does this in part by leading them into it through compelling everyday experiences. One might be inclined to assume that the virtue of these particular homey examples lies in their simplicity, but, on close inspection, that assumption proves to be incorrect; the analyses of the drunkard's walk and LIFE (the passive systems) are much less intricate and more easily followed than the analysis of batting averages in baseball.

The attractiveness and power of the opening examples derive, rather, from just the characteristic that will have to be deleted in Gould's final cosmic-evolutionary picture, namely, drivenness. The everyday examples that grab us and pull us along are laden, like those of mythic and epic adventure, with teleology: desire, will, striving, the call to battle. What we might term the Oprah factor – that is, the power of the personal, everyday struggle to induce empathetic, participatory involvement – is the factor that determines order of appearance in Gould's book, for "no experience could possibly be more intense than a long fight against a painful and supposedly incurable disease" (1997:45–6). Thus, phenomena that arise only in the last "micromoment" of cosmic evolution – battling disease, poker, baseball – go on stage first and dominate Gould's book until well past its mid-point.

Quotidian teleology draws upon and is drawn toward story. Early in *Full House*, Gould attributes our inclination to see trends where there are none to the fact that we are "story-telling creatures," for trends "tell stories by the basic device of imparting directionality to time" and make it possible for us to posit "a moral dimension to a sequence of events" (1997:30). From this point of view Gould's ultimate purpose is to dissuade us from seeing a story in biological evolution. But cosmic storylessness is the end of the journey; the beginning of the journey is Gould's discussion of his survival of disease, which he entitles "A Personal Story" (1997:45).[17] An example that is homey but not driven, such as the drunkard's walk, could not engage us in the same way. To be sure, the idea of randomness can be made to exert its own kind of appeal, as Leonard Mlodinow demonstrates in his recent *The Drunkard's Walk: How Randomness Rules Our Lives* (2008).[18] But this is a different sort of appeal than that projected in the agonistic, heroic tone cultivated by Gould in all of his lead-in examples, his championing of Darwin, and his own persona as a driven writer.

Unfortunately for Gould, any appeal to the agonistic impulse in order to make a non-teleological system memorable will run up against a conflict: the means of persuasion will at some level contradict the message. But Gould's strategy has by now been with us for some time – indeed, it seems, ever since certain thinkers became convinced that our human desires and aspirations lead us to distorted representations of the cosmos. Consider, for example, the first-century BCE Roman atomist Lucretius, champion of the Greek atomist Epicurus. Through his great epic, *On the Nature of the Universe*, Lucretius aims to convince the reader of the superiority of materialist explanation, that is, of explaining the workings of nature through random, impersonal "atoms" of matter rather than by reference to the thoughts, desires, and actions of wilful, human-like gods. Yet Lucretius punctuates his grand, poetically ensconced philosophical argument with the usual epic salutes to the gods of nature, and throughout analogizes Epicurus' voyage of mind to the treacherous, valorous odyssey of a sea-borne argonaut. Invocations of the workings of the gods in nature, as well as extravagant personification of nature, punctuate Lucretius' entire epic, even though his object is a dismissal of such views in favour of a materialistically conceived cosmos.[19]

Or, given Gould's interest in sports, we might hark back to an even earlier time and recall the famous "Achilles paradox" of the presocratic philosopher Zeno of Elea, who, like Gould, also sees in athletic contests an arena for abstract mathematical analysis. Zeno conjured up the spectacle of a race between two favourite story characters – Achilles of epic fame and the tortoise of fable fame (i.e., the race between the tortoise and the hare) – in order to entice his audience into a mathematical paradox. But, as in Gould's abortive fable, there is a self-defeating element, for the conclusion of Zeno's mathematical analysis – namely that motion is not real but an illusion – negates the spectacle of the race that draws the audience into the fable (and thus into the analysis) in the first place.[20] The task of leading non-specialists out of parochial cosmic visions seems to be perceived by Gould and his precursors as requiring an initial, robust indulging of precisely such parochial visions in order finally to dismiss them. The aftermath of seduction is abandonment.

In the course of *Full House* we experience extremes: a passionate involvement in the teleologies of will that arise in fighting disease and playing poker or baseball – and we might add, in doing science – giving way to a clearly non-teleological view of biological evolution. The change-up can even be seen in the fate of the book's title: invoked at first in the context of Gould's love of playing poker and his proposed wager with his readers, the phrase "full house" thereafter comes to refer instead to the mere fact of variation, whether driven or not.[21] The factor Gould wishes to disprove in the macroperspective is the same factor that will make the homey lead-ins, starting with his book's title, compelling. It's a cosmic bait-and-switch: Gould fields an opening lineup that he knows in advance will have to be sent to the showers before the final inning.

Gould is one of a kind; no other popular science writer so robustly infuses and enlivens the exposition of science with the passions of life, from quotidian to heroic – ingredients of the kind that engender and energize mythologies everywhere and give rise to the most intense micro-macrocosmic play. But thanks to the grand flip and the pattern of seduction and abandonment and their uncertain motivation, no other popular science writer leaves me with so many doubts as to whether such matters belong in the exposition of science.

EPILOGUE: BARROW AND GOULD

As any skilled cross-examiner knows, it is easy to find the truth one wants; not by inventing it, but by allowing the emergence of only that part of the whole truth that one wants to hear.

John Barrow, *The Artful Universe* (1995:246)

In many respects, Barrow's and Gould's analyses overlap. Most notably, both Barrow and Gould focus on wall arguments; and in both cases, these arguments interact in interesting ways with notions of contingency – or butterfly arguments. As is typical of popular science writers, both Barrow and Gould also challenge anthropocentric conceits: Barrow, our overestimation of our autonomy from the constraints of nature; Gould, our assumption that we represent the culmination of an evolutionary *telos*. But even while challenging such conceits, and once again in a move typical of such writers, Barrow and Gould both engage in anthropocentric persuasion, offering nominally scientific visions steeped in mythic appeal. This mythic appeal is most apparent in Barrow's elaborating upon a scenario of human ascendancy that we already know from mythology, and in Gould's leading us into his vision of our place in the cosmos through just those realms of intense human experience that offer alluring microcosms in many traditional mythologies. The mythic appeal is evident in many other ways as well, including some interesting resonances with the traditional fable format of a contest between members of two species of different physical stature, one large and the other small.

Moreover, both Barrow and Gould present themselves as connoisseurs of art as well as practitioners of science, each offering commentaries on the relationship between the two realms. For both scholars, though in different ways, wall reasoning facilitates passage between the two halves of the great academic divide, C.P. Snow's "two cultures." The humanities and humanities-leaning social sciences have long operated within an antinomy of, on one hand, rejecting the possibility of precise replication and prediction while, on the other, insisting upon the reality of principles of order and pattern in human thought and culture. For some, chaos theory embodies something like this antinomy, and thus offers a point of connection between the humanities and physical sciences.

There is also an important methodological convergence. Barrow and Gould both start with small, elementary examples that demonstrate the usefulness of wall reasoning in the mathematical representation of events defined by a limited set of variables; both then move to macro-deployments of their chosen mathematical devices – intersecting vectors for Barrow, the curve of distribution for Gould – and purport to tell us something about our place in cosmic evolution. In both cases, the macro-deployment involves an impressionistic stroke, in which the quantitative dimension of the analysis too becomes impressionistic in even the most basic sense (i.e., lacking in numbers). Within specific domains, wall reasoning may be rigorous, but applied to the realm of contingently conceived cosmic and biological evolution, wall reasoning, at least as practised by Gould and Barrow, is ad hoc and inadequately elaborated as a methodology.

Such methodological looseness opens the possibility of disparity and idiosyncrasy in end product, and for Barrow and Gould the divergence could not be more profound. Despite the many points of contact, Barrow and Gould offer us fundamentally different visions of the human place in the cosmos – visions that tug the heartstrings in opposite directions. Barrow's leaves us marvelling at our cosmic exceptionality, while Gould's leaves us crestfallen vis-à-vis other contenders for that special place (especially bacteria). The opposite pull of the two visions flows in part from basic choices made in implementing the logic of walls. Both Barrow and Gould emphasize walls defined by parameters of physical size, but these are implemented differently. Barrow presents a series of ever more specific walls that zero in on the capacity for culture as humans now know it: only our species (or something close to it) falls within the narrow range of size and other parameters that allow culture to emerge. By contrast, Gould sets up only one wall – the minimum size requirement for a living cell – so as to leave the human species maximally undistinguished vis-à-vis the other members of the set of living things. The diverging schemes are filled out impressionistically, with patches of science, eye-catching visuals, and a good deal of either Barrow or Gould. In both cases scientific observations are taken up within grander moral concerns – the kind that lead stargazers in many societies to overlay the constellations they observe with humanly potent lessons.

Finally, although the divergence between Barrow and Gould regarding our place in the cosmos is profound, it is not without irony, especially regarding the roles of necessity and chance. Barrow confirms a unique position for us as a result of our fitting necessary cosmic parameters, but this is tempered. Whatever its relation to his other works, the message that Barrow's *The Artful Universe* delivers is not that we had to occur (on this point there is not a complete departure from Gould), but rather that *if* a self-conscious, cultural being is to occur, it of necessity must be something like us. Gould emphasizes our cosmic unlikeliness with the expectation that recognition of our chance nature will deflate our pretensions; but some readers see in our very unlikeliness confirmation that we must be intended – as if, in matters of necessity and chance, the extreme of a position begets its opposite. This is certainly true as well of the other recent work, mentioned above, that seizes on the image of the drunkard's walk, Leonard Mlodinow's *The Drunkard's Walk: How Randomness Rules Our Lives*. In one strand of his argument, Mlodinow argues convincingly, indeed movingly, that understanding how thoroughly randomness figures in life can help us surmount the inclination to conclude, if we do not succeed, that we do not merit success. But taken to an extreme, an enhanced recognition of randomness in human affairs gives us one more reason to give up: it undercuts the conviction that virtue will be rewarded, which is a prime reason to keep trying – and a reason that, as in other matters of faith, can draw strength from naivety. Necessity and chance have always been baffling and paradoxical; no worldview, whether mythico-religious, philosophical, scientific, or everyday pragmatic, has succeeded in steering a calm passage through such matters.

∞

In chapter 2 I discussed Barrow's vision of our place in the cosmos and in chapter 3 I discussed Gould's largely opposite vision, in each case presenting reasons why I am unconvinced. Lest I be accused of sophistry, I will conclude with a comment about why I reject both of these nearly opposite conclusions about our place in the cosmos. The reason I reject them – or more precisely, the reason I reject either as science – lies in what they have in common: both offer limited findings spun into grand visions through impressionistic moves and the infusion of cultural preferences, political leanings, and humanistic sentiments (however noble these might be in themselves). To repeat a point made earlier: the quality of myth lies not in an indifference to empirical reality, but in readiness to leap from limited, local observations and findings to cosmic wisdom – a readiness that I find abundantly evident in both of these writers.

4

Copernican Kinship: An Origin Myth for the Category[1]

That quick dismissal of the idea of a universe without life was not so easy after Copernicus. He dethroned man from a central place in the scheme of things. His model of the motions of the planets and the Earth taught us to look at the world as machine. Out of that beginning has grown a science which at first sight seems to have no special platform for man, mind, or meaning. Man? Pure biochemistry! Mind? Memory modelable by electronic circuitry! Meaning? Why ask after that puzzling and intangible commodity? 'Sire', some today might rephrase Laplace's famous reply to Napoleon, 'I have no need of that concept.'

John Wheeler (Foreword to Barrow and Tipler, *The Anthropic Cosmological Principle*)

At rest, however, in the middle of everything is the sun. For in this most beautiful temple, who would place this lamp in another or better position than that from which it can light up the whole thing at the same time? For, the sun is not inappropriately called by some people the lantern of the universe, its mind by others, and its ruler by still others. [Hermes] the Thrice Greatest labels it a visible god, and Sophocles' Electra, the all-seeing. Thus indeed, as though seated on a royal throne, the sun governs the family of planets revolving around it. Moreover, the earth is not deprived of the moon's attendance. On the contrary, as Aristotle says in a work on animals, the moon has the closest kinship with the earth. Meanwhile the earth has intercourse with the sun, and is impregnated for its yearly parturition.

In this arrangement, therefore, we discover a marvelous symmetry of the universe, and an established harmonious linkage between the motion of the spheres and their size, such as can be found in no other way.

Copernicus, *De Revolutionibus*

Within the history of mythic visions allegedly shattered by science, the Copernican Revolution holds a special, epitomizing place. And so to catch Copernicus purveying solar myths is a mythologist's delight – a double delight, in fact, because there are, first of all, the wonderful solar myths that Copernicus rounds up for his reader; then, on top of these, the interesting friction that these create for our myth of Copernicus. Why, wasn't Copernicus the one who first led us to the vantage from which, for the first time, we could see the universe as it really is, and not as our petty, self-absorbed human-ity would have it? But, the sun a king? Planets as his children? The earth impregnated by the sun for her yearly parturition? This is the sort of thing that we expect from pre-Copernicans. As Thomas Kuhn has described the pre-Copernican world: "Though primitive concep-tions of the universe display considerable substantive variation, all are shaped primarily by terrestrial events, the events that impinge most immediately upon the designers of the systems. In such cos-mologies the heavens are merely sketched in to provide an enclosure for the earth, and they are peopled with and moved by mythical fig-ures" (1985:5).

It is possible, of course, to shrug this all off with the observation that it is certainly "nothing more than" metaphor, perhaps created by Copernicus as a mere sop for the mythically minded who would sit in judgment of his theory – though we must also ask whether this resolution does not sound a bit too contemporary and tidy.[2] Many comparative mythologists, especially it seems those of the nine-teenth-century social evolutionist school, were quick to argue that those who had left behind the age of myth still found artistic pleas-ure in the knowing use of mythical images as metaphors. The prom-inent nineteenth-century social evolutionist E.B. Tylor appealed to a formula to the effect that what is art for us today was literally held belief for archaic peoples. But might not the desire for such tidy dichotomization reveal a certain anxiety, if not regarding Coperni-cus' (and our own) metaphorical-mindedness, then at least regarding the literal-mindedness of the people of myth?[3] Even if we wished to hold out for a clear dichotomy, there is still much to examine in the very fact that certain metaphors are so broadly compelling. It would seem that the very reason that the kinship image of the universe occurring in Copernicus, the hero of science, has not trig-gered expressions of amazement is that no one perceives a prob-lem with the image; even for the scientifically minded, it is obvious,

natural, and little in need of analysis. Indeed, this highly mythologized vignette – depicting the sun as a lamp in a temple and as a ruler
surrounded by a family of planets – may well be Copernicus' most
quoted passage.

Copernicus' solar myths suggest many things, including, minimally,
that it is always possible to soften a scientific blow by – metaphorically, if you will – reinvesting the world with an anthropocentric
vision. Copernicus' familial images – kinship, kingship, parturition,
and so on – have a long history and call to mind numerous cross-
cultural analogues, for in many mythologies the sun is a parent or
ruler, other celestial bodies are kindred, and rituals of mating and
reproduction engender the large- as well as the small-scale rhythms
of the cosmos. If indeed myth is the nemesis of science, the situation
is especially ironic because the myths that Copernicus cites are, so
to speak, *the worst kind*: they are precisely the kind that nineteenth-
century "solar mythologists" attributed to the human mind in its
very infancy. The solar mythologists, a dominant voice in the study
of mythology in the second half of the nineteenth century, insisted
that all myths were ultimately inspired by celestial phenomena.[4] In
speaking of the first thoughts and words of archaic humanity, for
example, the German-born patriarch of this school, Friedrich Max
Müller, wrote:

> Every word, whether noun or verb, had still its full original
> power during the mythopoeic ages. Words were heavy and
> unwieldy. They said more than they ought to say, and hence,
> much of the strangeness of the mythological language, which we
> can only understand by watching the natural growth of speech.
> Where we speak of the sun following the dawn, the ancient poets
> could only speak and think of the Sun loving and embracing the
> Dawn. What is with us a sunset, was to them the Sun growing
> old, decaying, or dying. Our sunrise was to them the Night giv-
> ing birth to a brilliant child; and in the Spring they really saw the
> Sun of the Sky embracing the earth with a warm embrace, and
> showering treasures into the lap of nature. (1874:64)

There is yet another dimension of irony in Copernicus' words. Specifically, in the Judeo-Christian cosmogony of Genesis, we encounter not a sexually generated kinship cosmos, but rather a cosmos
created by a craftsman who stands over against his work, a relation

that is sometimes cited as a mythical prototype of the ideal of *objectivity* (even while anthropomorphically inflected as the "god's eye view"). Yet it is just this metaphor of objectivity that Copernicus' cosmic-family deposes. This all happens, of course, in parallel with the many other ways in which cosmic kinship, although denied in its officially sanctioned cosmogonic myth, finds its way into Judeo-Christian cosmology – from images of God as Father and Redeemer as Son, to the ideal of the nun as the consecrated bride of Christ, along with many other cults of folk religion that incorporate a sexually generative dimension in cosmology. One also recalls St Francis' famous *Hymn to the Sun*, which alternates between male and female siblings: Brother Sun, Sister Moon and Stars, Brother Wind and Air, Sister Water, Brother Fire, and Sister Earth, who is also referred to as mother.

The Creator in Copernicus' image is enmeshed within, rather than standing outside of, the cosmos as a would-be object of knowledge. We are left, then, with the irony that, in terms of images or emblems of objectivity, Copernicus' cosmic-family image is actually a step backward from the image it replaces, the divine Craftsman of Genesis. Or to put it differently: when epistemologists invoke the metaphor of the god's eye view, they do not seem to have in mind the god Zeus, with his perpetual struggles to find solace from his wrangling kinfolk, who are tutelary beings of the various regions of the cosmos. Zeus' great affliction as depicted by Homer – an affliction which the Judeo-Christian Creator had to some degree escaped, at least until Copernicus! – is, in a nutshell, that he cannot extricate himself from his cosmos, which is to say, from his family. In regard to our cosmic ambitions, Zeus and the Judeo-Christian God of Genesis form an interesting pair, reflecting, respectively, our conflicting desires to be enmeshed within and to stand contemplatively without our cosmos.

☙

In the two previous chapters, I considered works by a cosmologist and a paleontologist; the gaze of these works was outward, toward understanding the cosmos and the human place in it. In this chapter and the next, I consider, by contrast, works whose orientation is inward – toward understanding the workings of the human mind. These two chapters consider arguments from different fields within cognitive science, which George Lakoff, one of the thinkers to be considered, characterizes as "a broad discipline, covering everything from vision, memory, and attention to everyday reasoning and language" (1996:3); he identifies his own specialization, "cognitive linguistics," as the subfield of cognitive science "most concerned with issues of worldview, that is, with everyday conceptualization, reasoning, and language" (1996:3). My emphasis is theories of metaphor developed by Lakoff and his collaborator Mark Johnson – a topic that belongs to the more traditionally "humanistic" end of the cognitive science spectrum (the topic of the following chapter, artificial intelligence, will pull us toward the opposite end of that spectrum). Some readers who associate metaphor with humanities, art, and perhaps philosophy, may balk at science. It is important to recognize just how large the contemporary cognitive science umbrella is, and the energy and ambitiousness of ongoing attempts to bring traditionally humanistic topics under the protocols of science. Indeed, the study of moral reasoning from the standpoint of cognitive science (in some cases, specifically neuroscience), an enterprise which Lakoff and Johnson have helped to create, is one of the most noticeable growth areas of cognitive science, one which frequently spills over into the arena of popular science writing.

The specific topic from Lakoff and Johnson's work that figures in my analysis is kinship. Lakoff and Johnson's writings range from technical to popular in orientation, and the topic of kinship appears at different points within this range. At the technical extreme, kinship appears, in *Women, Fire, & Dangerous Things*, in Lakoff's analysis of philosopher Ludwig Wittgenstein's potent image of "family resemblance" posed as an alternative to the classic Aristotelian category (the contrast is elaborated in the final section of this chapter). In the middle of the continuum, kinship appears in emergent arguments about the genesis of moral reasoning, notably so in Lakoff and Johnson's *Philosophy in the Flesh*. At the maximally popularizing end of the continuum, kinship appears in Lakoff's recent political writings (e.g., *Don't Think of an Elephant!* and *Whose Freedom?*) in the form of a contrast between two models of family structure (the "Strict Father" and the "Nurturant Parent" models), along with the claim that such models can aid our understanding of the contemporary political world and our engagement in it. The technical theories lying behind kinship as moral metaphor are much simplified in these political works.

Since the scholarly enterprise examined in this chapter thus clearly contains an unabashedly popularizing arm, I should clarify a point about my choice of focus. Specifically, instead of Lakoff's recent political writings, I have chosen, as my main focus, *Philosophy in the Flesh*, a work that, compared with the political writings, is less obviously a popularizing platform, although it has been marketed to devotees of popular science writing, and, indeed, epitomizes at least one form of demographic ambition, that of academic interdisciplinarity. I have made this choice in part to call attention to a shift that takes place when Lakoff attempts to turn his theories about metaphor into a political message for the broadest possible audience. Lying between the highly specialized technical work and the work of mass appeal, *Philosophy in the Flesh* illustrates the route by which the one turns into the other. Specifically, *Philosophy in the Flesh* is a spawning ground for the ruminations on the relation of family structure and morality that infuse Lakoff's recent political writings. My choice of focus also allows for exploration of the influence of sociocultural context on category theory by way of a comparison with an earlier pair of theorists, French sociologists Emile Durkheim and Marcel Mauss, who took up similar intellectual problems in a different era, and who equally were impelled to take their academic message to the broader public in an effort to ameliorate the moral and political climate of their time.

While the following chapter deals with the dynamics of mental processing in the context of the ideal of artificial intelligence, the present chapter focuses on a basic unit often presumed to be implicated in such processing, i.e., the "category." Specifically, I consider the general idea of the difference between a mythological worldview and a scientific worldview from the standpoint of two topics – kinship and the category – that are often implicated in attempts to account for the transition from one worldview to the other, and that consequently serve as the site of numerous origin myths that sustain both.

MYTHOLOGICAL COSMOLOGIES OF KINSHIP AND OF THE BODY

The image of a cosmos organized and held together by kinship is found in many mythologies and/or religious traditions. Although my concern in this chapter is kinship cosmology in the "post-

mythological" world, it is important, as a preliminary step, to emphasize the complexity of those cosmologies of kinship that are typically
called mythological. To this end I will summarize some of the patterns that I have noted in research on the cosmological traditions of
Polynesia and especially of the New Zealand Maori. Those familiar
with the ancient Greek mythic cosmogony recounted by Hesiod in
his *Theogony* will note many parallels between this and the Maori
story.[5] Maori cosmology forms an epitomizing case for familial cosmology, since the Maori universe is conceptualized in most versions
as a kin-group. Perhaps the single most oft-quoted passage in the literature on the Maori is a passage from ethnographer Elsdon Best:
"When the Maori walked abroad, he was among his own kindred.
The trees around him were, like himself, the offspring of Tane [ances-
tral God of forests, and in some accounts, humans]; the birds, insects,
fish, stones, the very elements, were all kin of his ... Many a time,
when engaged in felling a tree in the forest, have I been accosted by
passing natives with such a remark as: '*Kei te raweke koe i to tipuna i
a Tane*' (You are meddling with your ancestor Tane)" (1924:128–9).
The specific characteristics and meanings of Maori kinship, as well
as the nature of the Maori commitment to the kinship cosmos (is this
a literal belief? a poetic trope?), pose complex issues; some of these
are pursued in the next chapter. In lieu of a complete analysis here, I
would like to offer some speculations about the way that this central
postulate – that the cosmos is a kin-group, the "kinship postulate"
for short – is expressed and constructed in Maori narrative.

First and most important, the kinship postulate seems to operate
as a kind of generative principle for a large group of etiological stories, of which there may be a theoretically limitless number. Maori
cosmogony begins as a version of the widespread story of the origin
of the universe from the mating of Sky and Earth, the first male and
female, as primal parents. In some Maori accounts the parents are
pressed closely together until the children separate them and then
go on to become fish, trees, plants, and humans and/or the tutelary
deities of these entities; they spread through the world as the variety of forms comprised in nature. Maori narratives often proceed
as though basing judgments of kinship descent more on functional
dependency (especially of smaller creatures on larger ones) than on
morphological similarity. The natural habitat of a given entity is
said to be its parent: shellfish are the children of the sea, birds the
children of trees, insects the children of plants. Thus a consistent

bio-cosmological principle – habitat as parent – stands behind the idea that Earth and Sky, the physical cosmos, are the parents of humans. While there are numerous forms of such interdependency, the web of kinship also comprises domination, jealousy, incompatibility, and, indeed, war. One can surmise that it is just because of the rich variety of kinds of relations – and the possibilities for multiple overlays of different, even contradictory, relations – found in kinship that the Maori have settled on it as their basic cosmological image, an image that finds a place for any entity.

One of the more discordant patterns displayed within these kinship stories has to do with the incompatibility of different life strategies displayed in the cosmos. The primordial example occurs in the story of the separation of Sky and Earth itself. In one version of this story, that recounted by the Arawa chief Te Rangikaheke,[6] one of the children is opposed to the idea of separating the parents. When his siblings overrule him and separate their parents, pushing Sky above, this dissenting child becomes Wind; with his brood of children, who are various forms of storms and pernicious meteorological conditions, he continually takes revenge on his siblings, including the humans, even to this day. In responding to the very first attack of Wind and his brood, there is dissension among the other siblings. Some think the sea would provide superior shelter, while others argue that greater security is to be gained on land. The argument ends in a stalemate: one group heads into the ocean and becomes the fish, the other group heads inland and becomes land dwellers. In splitting up to form the two tribes, the fish and the land dwellers both insultingly predict that the other will end up as food for humans. This scenario also allows a space for interstitial species – species difficult to classify. In the separation into land and sea creatures, one little band gets confused and heads in the wrong direction: these are the lizards, and thus we have an explanation of a particular land creature that bears many physical characteristics associated with the sea part of the family.

Countless other narratives reproduce the pattern of kinship disjunction, from stories about the mountains, which lived together until splitting up to occupy different tracts, to stories recounting the many finer differences that can be noted among insects. There is this story about how the mosquito and sandfly tribes split over a debate about strategy:

The sandfly said to the mosquito: "Let us go and attack man."
The mosquito remarked: "Let us await the shades of night ere we
attack, wait until evening, that we may hum in his ears and make
them tingle ... In this way we can safely attack, but if we attack
in daylight, many will perish. If you assail man by day, then you
of us two will surely be slain." Arose the sandfly: "Let us ever
give battle in the light of day. In daylight will I go forth. Though
I perish in myriads, what matter it, so long as I draw blood.
But you, O mosquito! Attacking by night, shall be destroyed by
smoke." Even so the mosquito consented, saying: "As you will."
(Best 1925:993)

One notes in this account the same pattern as found in the primordial separation of land and sea creatures: differentiation amidst
arguments and insults, speciation through incompatibility of life-
styles and distinct acts of will. These processes form sub-patterns
within the larger kinship postulate, and they also demonstrate the
intrinsic richness of this postulate – most especially in its capacity
to portray unity and disunity at the same moment and within the
same frame of reference. The accounts of the speciation of the natural world rely upon many of the idioms and much of the specific
terminology associated with human kinship and with processes of
tribal formation (notably the term *iwi*, tribe, for both human political units and natural species). Natural speciation provides a model
for human tribalization and vice versa.

But besides operating as a *generative* principle – of narratives, and
of models of social process – the kinship postulate also operates as a
principle of *incorporation*. The example most to the point here has
to do with the well-known European fable of the lazy ant and the
industrious grasshopper, a version of which appears in one of Elsdon
Best's collections of Maori narratives:

The ant spoke to the cicada saying: "Let us be diligent and collect much food during the warm season, that we may even retain
life when the cold season arrives." But the cicada replied: "Nay!
Let us rather bask in the sun's rays on the warm bark of trees."
Even so did the ant toil throughout the summer at gathering and
storing food in secure places. Meanwhile the cicada said: "What
a fine thing it is to bask in the warm sun and enjoy oneself. How

foolish is the ant who toils incessantly." But when the cold season came, and the warmth went out of the sun, the cicada perished miserably of cold and hunger, while the ant was warm and well fed in his snug home underground. (1925:991)

The European version of this story is a fable, that is, a terse story told to illustrate a specific moral. The Maori version is not a fable, but an origin myth, or part of one. The Maori version does not announce a specific moral; instead, like the story of the mosquito and sandfly and many other stories of the Maori, it accounts for and dramatizes the different behavioural characteristics of different species. The European story has thus been incorporated in such a way as to be consistent with other Maori stories detailing the unfolding web of Maori cosmic kinship.

I earlier alluded to Te Rangikaheke's account of the separation of the first parents, Sky and Earth, in which the dissenting child goes to the sky, becomes the wind, and with his meteorological brood makes war on the children who remain on the earth. The attacks occasion a series of debates among the earth-inhabiting siblings about the means of defence: some run to the sea (and become the fish); some burrow in the land (and become the plants). But there is an exception: the being who will become the progenitor of humans, Tū, refuses to be intimidated; he stands strong, rather than hiding. The battle between Tū (the human; the term literally means "Stand") and the elements ends in a stalemate, an apparent allusion to the present ongoing relationship between humans and "the elements." When this is achieved, Tū turns toward the siblings who, in seeking protection, had abandoned him and, in retribution, makes them his food. These events provide paradigmatic justification for humans eating the other things of the cosmos.

At points it is almost as if the truly significant units of Maori genealogies are the nodes of innovation and diversification in strategy which the above scenarios illustrate. At any rate it is important to note that these nodes, dialogical in form, provide a particularly open format for the incorporation of new negotiations, such as those involved in the dialogue of the ant and cicada. The deep and intense implication of the idiom of kinship in Maori narrative – in expounding both cosmic interconnection and diversification – is captured by distinguished scholar J. Prytz Johansen in a quaint image; he comments, "If one could picture to oneself a person like

KANT among the old Maoris – which indeed is difficult – one should not be surprised if to the fundamental categories of knowledge, time and space, he had added: kinship" (1954:9). I can think of no better analogy for describing how the kinship postulate operates as a generative and organizing principle for Maori narratives: it challenges the storyteller to create a story in broad conformity with other such stories; and, as a principle that is (to fill out the Kantian metaphor) a priori as well as synthetic, it confirms in advance that such a story is possible.

But there are other ways in which one could explore the relationship of the kinship postulate to narratives in the Maori case. One of these, briefly, consists in the fact that the kinship postulate has implications for the Maori organization of genres. There are Maori narratives that are reminiscent of familiar Western folkloric genres (myths, local legends, epics, proverbs, and so on). Yet cosmic kinship seems to add another level of integration, which links these narratives as one large narrative, because the characters from all of these narratives have a place in one encompassing cosmic genealogy, not totally unlike the ideal of "universal history" that has periodically appealed to Western historians. Rather than clear-cut myths versus historical legends, Maori narratives of the past flow into one another, beginning with something like myths and turning into something like legends – genealogy precluding absolute ruptures between genres.

The kinship postulate also has implications for the very concept of narrative. The genealogy that links the various narratives itself forms a narrative of sorts, so that a Maori cosmologist can tell the story of the origin of the universe and of human history merely by reciting a genealogy. For example, the first origins of the cosmos – the story of the children of Sky and Earth groping around in the dark between the yet-unseparated parents – are portrayed through a series of terms that has the appearance of a series of names in a genealogy; yet, in their literal meanings, the terms describe a series of sequential states of the early cosmos: Nothing, Night, Searching, Growth, and so on. They tell the same story as the narrative of the separation of Sky and Earth through a series of highly condensed allusions.[7]

෴

The foregoing provides one example of the elaboration of a kinship idiom in traditional cosmology, specifically that of the Maori; many other, highly varied examples can be found in the world. Before moving on to consider "categories," however, one more point should be added about cosmologies of kinship. Specifically, kinship, while designating a level of organization above that of any particular body, is yet ultimately inseparable from the idea of the body; bodies are the elements linked through webs of kinship, and indeed the body in its procreative power is the ultimate source of such webs. In the Maori case the interdependence of body and kinship as images of the cosmos is revealed, among other ways, in the fact that the first generative pair, Sky and Earth – sometimes with the implication that together they form a single body[8] – can be invoked to summarize the entire cosmos. Gaia plays a similar role in Greek mythology. Another example, indeed one of the best-known instances in Western cosmology, is the mythico-mathematical cosmos described by Plato in his *Timaeus* (32–4 [1981:44ff.]), in which the universe is characterized as a unitary living body. And these are not isolated instances. Although kinship provides what is probably the single most recurrent image or idiom through which mythologies portray the structure of the cosmos, the image of the cosmos as a body is not uncommon. Note, finally, that Copernicus (in the opening epigraph above) alludes approvingly to both images, referring to the sun as the mind of the universe (which is presumably the body) and the head of a royal family (with the planets as children).

PHILOSOPHIES OF FAMILIAR FORM: KINSHIP AND THE LIFE OF THE CATEGORY IN THE POST-MYTHOLOGICAL WORLD

The era of the presocratic philosophers, beginning around the sixth century BCE, is characterized in part by a critical confrontation of the mythological understanding of the world by methods and perspectives that came to be known as philosophy. And chief among the tools of philosophy is the category (a notion especially important to Aristotle and having a forerunner in Plato's notion of Forms). As a gross generalization, cosmic persons (gods and heroes and their genealogical interconnections) are to mythic understanding as categories are to philosophical understanding.

I will explore this contrast, emphasizing not the ways in which philosophy differs from myth, but rather the parallels and overlaps between the two. I do not disagree with the general proposition that new ways of understanding the cosmos appeared on the scene or at least rapidly picked up momentum in the era of the presocratics and classical Greek philosophers. But, in the company of many other scholars, I am also intrigued at the ways in which the transition is less than neat and complete (with some sympathy for the view that it will never be so). The love of illustration drawn from Hesiod and Homer (even amidst skepticism toward them) found in Plato and Aristotle, as well as myth-like formulations (such as the image noted above from Plato's *Timaeus*) are merely one early indication of the continuity of myth within the new philosophical spirit.

In this chapter I pursue the lack of completeness, in the transition from myth to philosophy, with respect to one particular theme. Specifically, in contemporary epistemology, and in certain theories of the category in particular, we encounter images of kinship and the body that cannot but recall the former life of these images in the world of mythology. I will focus here on the central if not centric engagement of kinship and the body within arguments about the very structure and nature of the category.

Although the term "category" now tends to be used mainly to denote a "class," or *group* of things classified together, the term was used by Aristotle (*Categories* [1938]) to designate ten ultimate predicates that summarized the possible attributes of things. Aristotle's categories are Substance, Quantity, Quality, Relation, Place, Time, Position, Condition, Action, and Affection. Aristotle seems to have arrived at his categories through an ascent in abstraction to the most general and encompassing possible predicates. The eighteenth-century philosopher Immanuel Kant drew on Aristotle's categories and introduced some modifications in order to construct his table of "*a priori* categories of the understanding." For Kant the categories constituted mental moulds through which reason unifies and imposes form on the data of sensory experience; by a priori he meant that the categories were prior to any such experience. Though Kant regarded knowledge as possible only through these categories, he rejected Aristotle's seeming conviction that the categories were transparent to being. According to Kant, the world we know is always and only the world as it is knowable to us through the a priori categories.

Aristotle and Kant tended to regard such categories as givens of consciousness, saying little about how humans came to possess them. One late nineteenth-century Darwinian response to the Aristotelian/Kantian categories, however, was to argue that they might be accounted for by bio-evolutionary process: adaptively advantageous mental moulds might be selected for by Darwinian processes – Herbert Spencer made such an argument.[9] Interestingly, John Barrow, whose *Artful Universe* we considered in chapter 2, also proposes (1995:28ff.) to Darwinize Kantian epistemology. This Darwinian response in one sense, of course, contradicts Kant by seeking categories that are a posteriori, i.e., products of experience over time. But evolutionarily advantageous categories would become part of the built-in mental structure that any particular human now inherits; in this sense the categories would "become" a priori. Kant had used the term "pure reason" to denote that which reason, by virtue of the a priori categories, is able to know prior to any sense experience. This new Darwinian project amounts to a search for an "impure" – that is, experiential – origin of "pure" reason.

The work of French comparative sociologists Emile Durkheim and Marcel Mauss in the early twentieth century also holds a place within the tradition of the search for the impure origin of the categories. And I argue that two influential contemporary scholars, George Lakoff and Mark Johnson, also represent the continuation of this tradition. My analysis focuses on a comparison of these two pairs of scholars: Durkheim and Mauss, and Lakoff and Johnson (for convenience sometimes shortened to Durkheim/Mauss, Lakoff/ Johnson). Both pairs pay homage to Aristotle and, though not adopting his formulations in detail, both ground their projects in part on the assumption that there exist basic, universal mental moulds something like those that Aristotle long ago set out.[10] Durkheim and Mauss find the source of the categories in the experience of *society*; Lakoff and Johnson find it in *bodily experience*. The Aristotelian/ Kantian categories are thus contested territory: the issue is, what kind of experience provides the model for the categories? The contesting claims will be the focus for the analysis that follows.

Before beginning, a terminological problem should be mentioned. Specifically, both Durkheim/Mauss and Lakoff/Johnson allude to the idea of categories in the Aristotelian/Kantian sense of very general mental moulds or predicates; and both choose as foci at least some of the categories set out by Aristotle and Kant. But both Durkheim/ Mauss and Lakoff/Johnson are interested in telling us also about the origin of "*the* category" in another sense, that is, in the more specific and contemporary sense of a "class" (or "set" or "kind"). Here the question is more specific: what is the source of our ability to create classes of things, i.e., to group together disparate entities into kinds or sets. It is all too easy to pass from one sense of category to the other, in part because the idea of class (or category, in our contemporary sense) is but a short step from some of the Aristotelian categories (notably Substance and Quality) and their Kantian derivatives, and in part because both senses of category are relevant to both Durkheim/Mauss and Lakoff/Johnson. But my purpose is to compare Durkheim/Mauss and Lakoff/Johnson in terms of the general drift of their arguments. For this purpose, the distinction between these two senses of category often is not critical; where it is I have tried to make the sense clear through the context.

THE ORIGIN OF CATEGORIES

Because many of the issues here are complex, I will summarize my arguments in advance, and then go on to fill them out by looking in more detail at the writings of Durkheim/Mauss and Lakoff/Johnson. My arguments can be summarized in three main points.

First, the image that informs Durkheim and Mauss' account of the origin of the categories is the image of kinship (either directly or as implicit in the idea of a "tribe" or "family"; Durkheim and Mauss regarded such kin-based units as the elementary forms of society). The image that informs Lakoff and Johnson's account of the origin of the categories is the human body. As indicated above, these two images – kinship and the body – are also two of the most recurrent images drawn upon in portrayals of the cosmos within mythology. My first main point is that the properties that make these two images good choices for mythmakers portraying the cosmos are the same properties that make them good choices for contemporary cognitive theorists seeking either the experiential origin of, or an origin myth for, categories or the category.[11] Both kinship and the body are ever-present in human experience. Both kinship and the body offer potent models of unity in diversity, of a whole made up of many interrelated parts. Images of kinship or the body in mythologies serve to organize and classify the cosmos; theorists attribute this same function to these images in accounting for the origin of the categories.

My second main point is that theories of the category offer as much potential for cosmological projection – that is, for imposing prosaic, local images onto the cosmos – as do traditional mythological portrayals of the cosmos. It is not difficult to recognize the process of such projection in a mythology that portrays the cosmos, as many in fact do, as one big tribe or as a gigantic living body. But, if Durkheim/Mauss are right, then we have all been unconsciously imposing on the cosmos the form of the tribe or family every time we recognize a set, for all sets must, at some level, bear the imprint of their original model. By contrast, if Lakoff and Johnson are right, then what our sets reflect is not the tribe but the body as container. If either the Durkheim/Mauss or Lakoff/Johnson theory of the origin of the categories is wrong (and it would seem that at least one of these theories has to be wrong, since they are contradictory), or if both theories are wrong, then the situation is even more interesting. Consider this: for those who have been convinced by either of these theories of the category, one of these images – the family or the body – takes on a cosmic stature just by virtue of this conviction. To argue that we are all necessarily projecting, through the intermediary of the category, a particular form on all cognized entities, ipso facto instantiates that image cosmologically by making it into an object of belief, though perhaps less obviously than do traditional origin myths when they project such images directly as the structure of the cosmos.

It is as if, in attempting to free ourselves from our parochializing inclinations, we forsake a mythological kinship-cosmos or body-cosmos for a cosmos based on a certain kind of abstract entity, the category, which meanwhile becomes the new object of our anthropocentric desires. In the tradition of Aristotle and Kant, categories are the enabling, mediating structures of all knowledge. A human image projected onto, or as, the category thus achieves, though by a less direct route, the same totalizing status as a human image that is painted by mythologies directly onto the canopy of the heavens.

My third main point is that juxtaposing two different theories of the impure origin of reason serves to bring to light the difficulty of adjudicating between contesting claimants. Both theories locate the source of the categories in those human experiences that they regard as most basic and formative. Opinions about what experiences are most basic and formative are academic discipline-centric. Durkheim and Mauss, who were attempting to forge a discipline of "comparative sociology," saw experiences of social collectivity as the linchpin; Lakoff and Johnson, who are developing a more psychologically oriented cognitive perspective see the experience of the body as the linchpin. Both pairs of scholars try to substantiate their claims by showing that the categories conform in outline to the experiences they posit as formative. The problem is that the categories, as predicates or moulds of maximal generality, conform in outline to virtually *all* experience, making it difficult to adjudicate between different claims to primacy. The deep differences in portrayals of the universal condition of all knowledge given to us respectively by Durkheim/Mauss and Lakoff/Johnson originate in different kinds of centric inclinations applied to the Aristotelian categories; by the former, the categories are sociomorphized; by the latter, they are (in the narrow sense of the term) anthropomorphized.[12]

Viewing mythological or folk portrayals of the cosmos as projections of local frames of understanding and practice is a rather standard anthropological move. Even the most vocally anti-positivistic of contemporary social theorists will often quietly accede to the Durkheimian/Freudian dictum that, whatever its professed object, the "real" object of mythico-religious cosmological discourse is humans and their inner-worldly concerns. Part of the force with which the cosmological "projectionist" thesis took hold within ethnography at the turn of the century had to do with the fact that amidst intense opposition on many other points, psychoanalytic schools (deriving especially from Freud) and sociological schools (deriving especially from Durkheim) were in fundamental agreement about where the true object of cosmological discourse is *not* to be found, namely, the heavens. For Freud the foundational experience was the generic individual psyche as situated within the dynamics of familial relationships, especially the relation of parent and child;[13] for Durkheim it was the experience of the social collectivity through its pre-eminent structures, symbols, and rituals. Freud and Durkheim together form a sort of Copernican Revolution within ethnology, in the sense that they charge this enterprise with laying bare the all-too-human desires and designs that impinge on cosmological discourse.

☙

Now to the specifics: I compare what Durkheim and Mauss, and Lakoff and Johnson have to say regarding the origin of the categories. Since both pairs of scholars have written a great deal, my discussion is selective, aiming to bring out the general tenor of both positions as well as some pertinent sub-themes within each.

Durkheim and Mauss' perspective can be found in two condensed statements from Durkheim's *Elementary Forms of the Religious Life*:

> At the roots of all our judgments there are a certain number of essential ideas which dominate all our intellectual life; they are what philosophers since Aristotle have called the categories of the understanding: ideas of time, space, class, number, cause, substance, personality, etc. They correspond to the most universal properties of things. They are like the solid frame which encloses all thought; this does not seem to be able to liberate itself from them without destroying itself. (1965:21–2)

How does Durkheim explain the origin of these categories?

> The category of class was at first indistinct from the concept of the human group; it is the rhythm of social life which is at the basis of the category of time; the territory occupied by the society furnished the material for the category of space; it is the collective force which was the prototype of the concept of efficient force, an essential element in the category of causality. However, the categories are not made to be applied only to the social realm; they reach out to all reality. (1965:488)

Let us turn for a comparison to Lakoff and Johnson; for present purposes it is most useful to focus on two works, Johnson's *The Body in the Mind* (1987) and Lakoff and Johnson's *Philosophy in the Flesh* (1999). These works present an account of the origin of categories that is largely devoid of reference to the sort of social or collective forms that play the dominant role in Durkheim and Mauss (although, as we shall see, the family makes an appearance when Lakoff and Johnson turn from epistemology proper to moral philosophy).

Lakoff and Johnson employ the concept of "image schemata," defined by Johnson as "a recurrent pattern, shape, and regularity" that arises "chiefly at the level of our bodily movements through space, our manipulation of objects, and our perceptual interactions" (1987:29). As "*structures for organizing* our experience and comprehension" (1987:29), the function of the image schemata is roughly comparable with the functions that Aristotle and Kant, and Durkheim and Mauss following them, attributed to the categories.

And, more importantly, some of the specific image schemata that Lakoff and Johnson focus upon are just those potential predicates that Aristotle, Kant, and Durkheim and Mauss call categories. To facilitate comparison I focus on three such categories: class, cause, and time – in each case filling out Durkheim and Mauss' arguments, then comparing them with the arguments of Lakoff and Johnson.

Class

Durkheim and Mauss' most elaborate line of argument has to do with what one might call the morphological congruence between the idea of "society" and the idea of a "class." One of the ways, they argue, in which any class of things is like a society is that one perceives in any class some sort of internal cohesion, an affinity among the members of that class. The loftiest moments in Durkheim and Mauss' writings occur when they address the issue of social sentiment, the collective "effervescence" that affectively binds together members of society. Such sentiments reach peaks of intensity in social rituals, fusing individuals into a single entity "this moral being, the group" (Durkheim 1965:254). The experience of this all-embracing, all-powerful sentiment – this absolute basis of society – is seen by Durkheim and Mauss as the experiential model for the internal cohesion we ascribe to any logical class of things.

Another way, for Durkheim and Mauss, in which any class of things is morphologically congruent with a society is that any class, like any society, implies a border. In an intriguing argument, Durkheim and Mauss point out that although logic has nothing per se to do with space, we nonetheless tend to diagram logical relations spatially (i.e., with what are now called Venn diagrams). Logical diagrams often involve circles within circles, just as tribal structures often involve sub-clans within clans. This proclivity suggests to Durkheim and Mauss that the tribal territory, border, and concentric structure is the experiential source of the space, border, and concentric structure of formal logic. "It is certainly not without cause that concepts and their interrelations have so often been represented by concentric and eccentric circles, interior and exterior to each other, etc. Might it not be that this tendency to imagine purely logical groupings in a form contrasting so much with their true nature originated in the fact that at first they were conceived in the form of social groups occupying, consequently, definite positions in space? And have we not in fact seen this spatial localization of genus and species in a fairly large number of very different societies?" (1972:83). They summarize: "Logical relations are thus, in a sense, domestic relations" (1972:84).

Let us turn now to see what Lakoff and Johnson have to say about the origin of the category of class. If Durkheim and Mauss find the model for the idea of any clearly bounded class of things in the formative experience of being a member of a tribe or family, Lakoff and Johnson find it in the experience of one's body – specifically of one's body as a container or as an entity that deals with other containers. This bodily experience gives rise to what they call the CONTAINER schema. The formulations concerning this schema that offer the most dramatic contrast with Durkheim and Mauss are to be found in Johnson's *The Body in the Mind*; Johnson writes:

> Our encounter with containment and boundedness is one of
> the most pervasive features of our bodily experience. We are
> intimately aware of our bodies as three-dimensional contain-
> ers into which we put certain things (food, water, air) and out of
> which other things emerge (food and water wastes, air, blood,
> etc.). From the beginning, we experience constant physical con-
> tainment in our surroundings (those things that envelop us).
> We move in and out of rooms, clothes, vehicles, and numerous
> kinds of bounded spaces. We manipulate objects, placing them in
> containers (cups, boxes, cans, bags, etc.). In each of these cases
> there are repeatable spatial and temporal organizations. In other
> words, there are typical schemata for physical containment.
> (1987:21)

Paralleling the quest of Durkheim and Mauss, Johnson pushes further, claiming that the very principles of formal logic derive from just this bodily experience. His arguments, like those of Durkheim and Mauss, appeal mainly to morphological congruence. For example, Johnson argues that the P/~P formula of formal logic is congruent with the CONTAINER schema that we derive from the bodily experience of boundedness; thus bodily experience is the likely source of formal logic, which extends and formalizes it. "It follows from the nature of the CONTAINER schema (which marks off a bounded mental space) that something is either *in* or *out* of the container in typical cases. And, if we understand categories metaphorically as containers (where a thing falls within the container, or it does not), then we have the claim that everything is either P (in the category-container) or not-P (outside the container). In logic, this is known as the 'Law of the Excluded Middle'" (1987:39).

In *Philosophy in the Flesh*, Lakoff and Johnson add that the CONTAINER schema forms the basis of the philosophical notion of "essence"; their argument, like that of Durkheim and Mauss, implicates the space, border, and concentric structure of formal logic.[14] "For the sake of imposing sharp distinctions, we develop what might be called *essence prototypes*, which conceptualize categories as if they were sharply defined and minimally distinguished from one another. When we conceptualize categories in this way, we often envision them using a spatial metaphor, as if they were containers, with an interior, an exterior, and a boundary. When we conceptualize categories as containers, we also impose complex hierarchical systems on them, with some category-containers inside other category-containers" (1999:20). In *Philosophy in the Flesh*, we encounter a further interesting twist in the appeal to morphological congruence. In this ambitious work, Lakoff and Johnson aspire to reconceptualize not just logic, but the history of philosophy from the standpoint of "the striking claim that the very structure of reason itself comes from the details of our embodiment" (1999:4). They attempt to do this by revealing the image schemata at the basis of various philosophical doctrines. Aristotelian logic and theory of essences, they argue, are governed by the CONTAINER schema; their presentation begins, "Aristotle gave us the classical formulation of what we will call 'container logic,' which arises from the commonplace metaphor that Categories Are Containers for their members" (1999:380). This is a novel and intriguing attempt at substantiating claims about the experiential source of logic. That is, in addition to demonstrating the morphological congruence between the experience of the body and the operations of logic, Lakoff and Johnson attempt to demonstrate morphological congruence by means of a reading, from their own perspective, of the philosopher credited with the invention of formal logic.

To summarize: Durkheim and Mauss give us the in-the-tribe/outside-the-tribe origin of logic; Lakoff and Johnson give us the in-the-body/outside-the-body origin.

Cause

In summarizing Durkheim and Mauss' views on the origin of the category of "cause," I can do no better than to repeat a part of the long passage quoted above: "it is the collective force which was the prototype of the concept of efficient force, an essential element in the category of causality" (Durkheim 1965:488). "Collective force" is the power of the group over the individual. For Durkheim and Mauss this power manifests itself most efficaciously not through physical coercion but through symbols and ideas that create and arouse a sui generis energy and sense of moral authority within a social collectivity – a dimension of consciousness that transcends individual interests. It is the feeling of the power of social sentiment and moral authority that is the experiential source of the idea of cause, or the notion that the things of the cosmos are not isolates but rather exert influence on one another.

Let us now consider Lakoff and Johnson on the category of cause, which, paralleling Durkheim and Mauss, they refer to as force, or, more specifically, the image schema of FORCE. Contrary to Durkheim and Mauss' affective, emotional, moral force emanating from the collective, Lakoff and Johnson find the experiential source of force in physical force. Johnson writes: "Though we forget it so easily, the meaning of 'physical force' depends on publicly shared meaning structures that emerge from our *bodily experience* of force. We begin to grasp the meaning of physical force from the day we are born (or even before). We have bodies that are acted upon by 'external' and 'internal' forces such as gravity, light, heat, wind, bodily processes, and the obtrusion of other physical objects. Such interactions constitute our first encounters with forces, and they reveal patterned recurring relations between ourselves and our environment" (1987:13).

The phrase "publicly shared meaning structures," near the beginning of this passage, offers an opportunity to clarify a point of potential confusion between the theories considered here. One does find numerous references to a social dimension for the image schemata throughout the writings of Lakoff and Johnson. But it is important to realize that the social dimension enters the picture for them in a very different way than it does for Durkheim and Mauss. For Durkheim and Mauss the formative experiences that give rise to the categories are all intrinsically social: they are various dimensions of the irreducible experience of social collectivity. For Lakoff and Johnson the formative experiences belong to the body: there is nothing to suggest that a body in isolation from society would not come up with the image schemata discussed above. Although, practically speaking, it would seem necessary that there be a second body to care for the infant body undergoing the formative experiences, the experience of a social relationship with that second body is not portrayed as otherwise intrinsically necessary for generation of the image schemata. The category of class, for example, derives pre-eminently from the experience of the first body in itself as a container; along with this are mentioned environmental interactions that apparently need not be with animate beings. A necessarily social dimension enters Lakoff and Johnson's perspective not on the level of formative experience, but in the claim that at least a part of embodied experience is universal: it is the universality of bodily experience that gives rise to publicly shared meaning structures.

Lakoff and Johnson's main discussion of cause in *Philosophy in the Flesh* is found in chapter 11 ("Events and Causes") (1999:170–234): they address the issue of whether the multiple senses of cause yield an underlying unity; and they answer in the affirmative. In the opening part of their presentation they say: "At the heart of causation is its most fundamental case: the manipulation of objects by force, the volitional use of bodily force to change something physically by direct contact in one's immediate environment. It is conscious volitional human agency acting via direct physical force that is at the centre of our concept of causation ... Prototypical causation is the direct application of force resulting in motion or other physical change" (1999:177). In closing their discussion of cause, they provide a final terse example: "The central prototypical case in our basic-level experience gives us no problem in answering the question. He punched me in the arm. He caused me pain. Yes, causation exists" (1999:233).

To summarize these theories of the origin of the category of cause: for Lakoff and Johnson, the origin lies in the experience of physical force exerted by or against the body; for Durkheim and Mauss, the origin lies in the sense of a *sui generis* transcendent energy or power felt in collective milieux, eminently in social ritual.

Time

Argument by morphological congruity takes an interesting turn in both Durkheim and Mauss and Lakoff and Johnson, in both cases involving an appeal to the reader to engage in introspection.

Compare the following two passages on the category of time, the first by Durkheim, the second from Lakoff and Johnson:

Durkheim:

> Try to represent what the notion of time would be without the processes by which we divide it, measure it or express it with objective signs, a time which is not a succession of years, months, weeks, days and hours! This is something nearly unthinkable. We cannot conceive of time, except on condition of distinguishing its different moments. (1965:22)

Lakoff and Johnson:

> Try to think about time without thinking about whether it will *run out* or if you can *budget* it or are *wasting* it.
>   We have found that we cannot think (much less talk) about time without those metaphors. That leads us to believe that we conceptualize time using those metaphors ... What, after all, would time be without flow, without time going by, without the future approaching? (1999:166)

The two accounts are so parallel that one wonders which is the case: that Lakoff and Johnson have unwittingly reinvented Durkheim's rhetoric, or that the former regard the latter as so broadly known that readers can be trusted to recognize the allusion. Consider what happens in these parallel accounts. In each, the appeal to the inconceivability of purely abstract time leads the analysts to propose a more concretized portrayal – of measured or differentiated time – but in each case with a selected version of concretized time, one that supports the larger claims of the particular authors. Durkheim slips in the calendar, that is, socially constituted units such as weeks and days, which lead directly into his overriding concern with social/ sacred ritual; whereas Lakoff and Johnson slip in metaphors and conceptualizations ("budgeting" of time, "flow"), whose primary contexts are typically a generic individual negotiating quotidian life. The most abstract and fluid of all concepts, time is entirely congruent with either concretization.

HOW IT ALL BEGAN

Continuing with the temporal theme, but at another level of analysis, it is important to note that, in arguing for their respective versions of experiential origins of the category, Durkheim and Mauss and Lakoff and Johnson both focus on the morphological congruity between the structure of the category, on one hand, and, on the other, their favoured realm of experience: the social collective for Durkheim and Mauss, the generic body for Lakoff and Johnson. But these structural arguments lead into theories of origin that invoke or allude to a variety of temporal or quasi-temporal frameworks, including the individual's psychic ontogeny, the history of Western academic thought, the social evolution of human institutions, and the biological evolution of the human species. Often several temporal frameworks are alluded to at once, with a suggestion of the recapitulation of similar patterns within different time scales. These quests for temporal founding moments sometimes have the feel of a quest for an origin myth. As if to make up for not offering a full-blown, heroic origin myth of the kind we find in Hesiod or the Maori, the formulations of Durkheim/Mauss and Lakoff/Johnson are myth-like in two different ways: first, as discussed above, in that they project, by mediation of the category, a particular local image onto the cosmos; and second, in that they attempt to give us some version, even if vague, of a temporal moment of origin, when that particular image came to be so favoured.

For example, along with overtures to Darwinian biological evolutionism, one encounters in Lakoff and Johnson allusions to ontogenic development; some of the clearest occur in the passages from Johnson's *The Body in the Mind* that were cited above: "We begin to grasp the meaning of physical force from the day we are born (or even before)" (1987:13) and "From the beginning, we experience constant physical containment in our surroundings (those things that envelop us)" (1987:21).

But just what is our experience of force and envelopment before we are born, i.e., the experience of the womb? Is it the body's first experience of itself as a container (perhaps a container interacting with another container)? Or does the womb provide the first experience of genealogy, encompassment by tribal territory, and the individual's dependence on the social? Such appeals have mythical dimensions; they attempt to construct a plausible first experience and are necessarily speculative. Even "empirical" research into child development cannot escape the necessity of inferring the experience of those who are as yet unable to articulate it. Throughout the history of social thought, the minds of infants have furnished an irresistible object of learned projection, an ontogenic alternative to the mythical/historical first caveman stepping out of the cave. The fact that infants are like us yet different – ourselves at a distant, less-developed stage – together with their limited capacity of responding (at least on such matters as the relative priority of physical vs. moral experience) places them in the company of other favoured objects of projection, including ancient ancestors, culturally exotic "others," non-human animals, and celestial bodies. Durkheim and Mauss adopted the framework provided by the socio-cultural evolutionism of their day as their arena for speculating on category origins; and this framework, too, was inclined toward the position that the moral and intellectual development of the individual recapitulates human socio-cultural evolution.

In assessing the possible arguments for one view or the other, it is important to mention an asymmetry that might seem to favour the Lakoff/Johnson position. Sometimes Lakoff and Johnson are, and sometimes they are not, careful to distinguish the several possible senses of their mantra, "embodied." Even among committed Cartesian dualists I suspect one could find some who would assent to the proposition that all cognition is embodied *in some sense*. To the extent that one accepts that the site and mechanism of the categories must be physical/neural structures, the categories must be embodied; and in *Philosophy in the Flesh* and other works Lakoff and Johnson offer glimpses into contemporary neural theory and how it might relate to their theories of metaphor—although the relation is presently so distant that the former realm of theory sheds little light on the latter.[15] Whether the category-shaping experience is the body or society, this experience can be encoded only in the neural structures of the body, because there is no corresponding unitary neurological entity – no body in this physical sense – that belongs to society as a whole. I accept this point as indisputable, and my reading of Durkheim and Mauss suggests that they, too, would assent to it. But from the fact that the body is the organ of encoding and contains the means of encoding, it does not necessarily follow that the body is the model or form encoded.

If Lakoff and Johnson favour ontogenic appeal and Durkheim and Mauss cultural-evolutionary appeal, they converge in the dream of utilizing their respective theories about the origin of categories to put all of academia on a new footing – a more modern version of the ambition encapsulated in Kant's famously arrogant title, *Prolegomena to any Future Metaphysic*. Not only will they show the necessity of their perspective, they will enshrine it at the base of the intellectual world, structurally and temporally. The crown of *regina scientiae* will go to the discipline that succeeds in accounting for the origin of the categories.

Both pairs – Durkheim and Mauss, and Lakoff and Johnson – specifically focus on philosophy, acknowledging the historical prestige of this discipline while challenging what they see as an inclination toward a view of reason as "pure." Both pairs of contenders seem to think that Aristotle got the categories formally about right, and that what is lacking is a proper accounting for their origin and development. Both Durkheim/Mauss and Lakoff/Johnson also envision a major rewriting of the history of philosophy from their perspectives, a rewriting that begins with a new account of the origin and nature of the fundamental categories.

For Durkheim and Mauss, the Aristotelian categories offered an organizational plan for comparative sociology. The history of the Aristotelian categories would be cast as the continuation of a sort of prehistory of the categories, in which each Aristotelian category would be seen to be prefigured in the social forms and practices of tribal societies. From tribal classifications and practices it would be possible to reconstitute the social ground and prehistory of the category of class; from food customs, the category of substance; from exchange customs the category of relation; and so on.[16] Calling attention to perduring questions about the origin of Aristotelian categories, Durkheim and Mauss conclude their investigation this way: "As soon as they are posed in sociological terms, all these questions, so long debated by metaphysicians and psychologists, will at last be liberated from the tautologies in which they have languished. At least, this is a new way which deserves to be tried" (1972:88).

The social evolutionism that undergirds Durkheim and Mauss' plan – in which contemporary tribal societies are thought to offer something like the prehistory of the Aristotelian categories – is not found in Lakoff and Johnson; indeed such social evolutionist doctrines have by now been generally discarded. Yet Lakoff and Johnson's plan is in many respects parallel to Durkheim and Mauss': Lakoff and Johnson seek the basic character of the Aristotelian categories in non-technical, everyday understandings of the world, as revealed in contemporary popular metaphors. Lakoff and Johnson then analyze the history of such categories in Western philosophy as a continuation and special case of such everyday understandings. They account for different philosophical schools through the different embodied image schemata and metaphors that these schools choose to emphasize – in a word, this is the project of Lakoff and Johnson's *Philosophy in the Flesh*.

In sum, both Lakoff/Johnson and Durkheim/Mauss aspire to rescue and reframe Western philosophy by reconceptualizing its root doctrines from the standpoint of their respective versions of the experiential origins of reason, emphasizing especially the great founding figure of Aristotle.

୧୬

These two projects – Durkheim/Mauss' and Lakoff/Johnson's – overlap in opposing the view of philosophy as a pure, self-contained discourse, unencumbered by the constraints of the mundane world. That is, Durkheim/Mauss and Lakoff/Johnson both insist that the basic conceptual apparatus of higher-level thought – philosophy, logic, science, and math – grow out of, and never entirely sever connections with, everyday knowledge and practice – inflected, by Durkheim and Mauss, as our pre-scientific, tribal past, and by Lakoff and Johnson, as contemporary everyday understandings of our life in the world. It is important to note that at this level Durkheim/Mauss and Lakoff/Johnson are allied against the idea of pure reason.

But there is not just one form of everyday understanding. We see a definite selectivity operating in the two projects, even though both choose data that happen to be of special interest for mythologists and folklorists. For their examples of pre-scientific, everyday understandings, Durkheim and Mauss draw on *myths* and *rituals* from tribal societies, which they considered to be models of coherent social integration; Lakoff and Johnson emphasize *everyday metaphors* (many of which have a proverb-like flavour) primarily

drawn from their own society. Both in terms of the kind of societies involved ("tribal" in contrast to "modern") and of the speech
genres that are considered (myth and ritual in contrast to everyday
metaphor), there is a link between what the scholars would like to
show (collectivity in contrast to the body) and the kinds of everyday
understandings they have selected as their data. Myths and rituals,
the data that Durkheim and Mauss draw upon, have a social orientation and functionality: they portray and uphold the overall worldview and social structure of the societies in question. By contrast,
Lakoff and Johnson, in surveying the range of metaphors, particularly emphasize everyday phrases and proverbs or proverbial sayings
of the kind relevant to any individual's negotiation of the everyday
world in pursuit of specific purposes: time is money, budget one's
time, life is a journey, and so on. It is worth noting that studies by
folklorists of proverb and proverbial speech typically call attention
to the pragmatic character of these folk genres; though proverbs are
social in the sense of being broadly disseminated among members of
a society or language group, their advice is typically directed to, and
consulted by, an individual facing a concrete dilemma.

❧

There is of course a major quirk in all of this, specifically that many
images can be used to reference either the social collective or the generic individual, depending upon how one reads them. For example,
Lakoff and Johnson, in their broad survey of metaphors, come
upon instances of the metaphor whose source image for Durkheim
and Mauss served as exemplar of the collective – specifically, the
family and kinship. But when this happens, Lakoff and Johnson
invariably read these in terms of the generic body. The most striking example occurs in chapter 14 ("Morality") of *Philosophy in the
Flesh*. Here, Lakoff and Johnson turn to the metaphor of the family
as the overarching metaphor that binds particular moral metaphors
into a "coherent moral view" (1999:312), asserting that this metaphor, enlarged to "The Family of Man," provides the basis of universal morality. "Since the metaphor projects family moral structure
onto a universal moral structure, the moral obligations toward
family members are transformed into universal moral obligations
toward all human beings" (1999:317). In this passage, the origin
of universal morality is said to lie in moral obligations that belong

to family structure. The passage is highly reminiscent of Durkheim and Mauss, positing the experience of the immediate social collective as the ground of universal moral order, which is projected outward from the first and primary experience of moral structure (the family for Lakoff and Johnson; the tribe for Durkheim and Mauss). At this point, Durkheim and Mauss see the matter as finished, but Lakoff and Johnson do not, for their larger perspective demands that the body be the generative foundation, not just for categories but also for morality, whether local or universal. For them, universal morality reduces in its principle to family morality, which in turn must be subjected to another reduction, which they attempt in the following passage: "Our very idea of what morality is comes from those systems of metaphors that are grounded in and constrained by our experience of physical well-being and functioning. This means that our moral concepts are not arbitrary and unconstrained. It also means that we cannot just make up moral concepts *de novo*. On the contrary, they are inextricably tied to our embodied experience of well-being: health, strength, wealth, purity, control, nurturance, empathy, and so forth" (1999:331).[17]

Note that this list proceeds from qualities most easily attributed to an individual body (health, strength) to those involving a relation between bodies (nurturance, empathy). This ordering is the equivalent, in the realm of morality, to the ordering of physical experiences offered in the passage quoted above in which Johnson offers "our bodies as three-dimensional containers" as his *first* example of the embodied experience of containment, and then moves outward to offer other examples of embodiment that involve our relationships with things outside of ourselves, such as vehicles and other containers. In both realms – morality and physical experience – we start with a bio-psychological sense of body as sentient organism and "self" and we move outward into a world of bodies in interaction. Apparently, either end of such continua counts as "embodied."

❧

But the fact that what is entailed in "embodiment" is so stretchable poses a vexing point about metaphor in general, and about the relation of Lakoff and Johnson to Durkheim and Mauss. Specifically, one with the will to do so can bring virtually any two terms into a metaphorical relation: metaphorizing means making a choice – in many cases, a political choice – from a pool of possibilities. We certainly could, in the manner of the examples above, extend the notion of *embodied* experience to include the experience of containment within the tribal territory. And if embodied experience includes such qualities as nurturance and empathy, why not say that the experience of collective effervescence in ritual, which for Durkheim epitomizes the irreducibly "collective," is an embodied experience? By this route we could arrive at the conclusion that the difference between Lakoff/Johnson and Durkheim/Mauss is non-existent. But such a conclusion is contrived: the very ordering of Lakoff and Johnson's litanies suggests a hierarchy of what counts as embodied. The body's awareness of itself as container, or of its own strength and health, are offered first; however far the notion of embodied experience be stretched, the stretching is outward from an anchored beginning point. The founding and final message that comes through, at least when one considers the full corpus of Lakoff and Johnson's work is: *the body.*[18] The anchoring point and final message in Durkheim and Mauss is: *society.*

Yet it is interesting to note how it is possible to start with a certain frame, the body for Lakoff and Johnson, and draw in other frames, such as the collective dimension of kinship. Since the body and its genealogy are mutually implicating, analytical passage from one to the other is always possible – no less in theorizing about the category than in traditional mythology: recall the Maori mythic cosmos described above, in which the cosmos is portrayed at different moments as both a genealogy and a unitary body. To Lakoff and Johnson's referral of kinship to the body, moreover, one finds a fascinating reciprocal in one of Durkheim's students, who finds a way to refer the overtly bodily to the irreducibly collective. Specifically, Durkheim's student Robert Hertz adopted the same imperative embraced by Durkheim's other students: to reveal the collective nature of human life and thought. But unlike those other students, who chose topics with plainly evident collective dimensions – such as social rituals and forms of ritualized exchange – Hertz landed a topic whose immediate referent is the body as physical organ: the asymmetry that favours one side of the body over the other (in humans usually the right over the left). To the given physical asymmetry of right and left, Hertz argued, nearly every society has added a collective symbolic dimension, in the form of connotations of sacred for the right and profane for the left (obvious examples are

visible in derivatives of the Latin terms *dexter* and *sinister*). That
the body is given in nature as asymmetric is beyond human volition; but that this given condition of the body is universally seized
upon as a moral symbol is a matter of collective sentiment. Hertz
suggests that "if organic asymmetry had not existed, it would have
had to be invented" (1973:10).[19] Thus, we see in Hertz's reading
of the social collective in the body something like the reciprocal
to Lakoff and Johnson's reading of the body in the genealogy. If
(as noted above) one could deny the differences between Lakoff/
Johnson and Durkheim/Mauss only at the cost of considerable con-
trivance, it must also be acknowledged that considerable contriv-
ance went into the creation of those differences.

Creative incorporation in both directions – toward the body in
Lakoff and Johnson; toward the moral collective in Durkheim and
his students – is a fascinating business, illustrating the complexity
one encounters in these heady claims about the ultimate ground of
knowledge. In the end there may be no image that is irreducibly
collective or bodily, yet images can be tweaked to conform to analytical – and, of course, political – aspirations. What comes through
respectively in Durkheim and Mauss, and Lakoff and Johnson, are
very different origin myths of the category, which spill over into different theories and rhetorics of moral obligation.

There are a number of interesting ways in which the worlds of
Durkheim and Mauss and Lakoff and Johnson permeate their origin myths of the category. "He punched me in the arm. He caused
me pain. Yes, causation exists." Although Lakoff and Johnson's
little scenario falls well short of Freud's famous primal patricide-
cum-totemic meal (*Totem and Taboo*) – an act of violence which,
for Freud, marked the origin of human consciousness as we now
experience it[20] – it nonetheless offers an interesting counterpoint to
Durkheim's scenario of the category of cause arising out of cohesive
social sentiment. Construed as the shaping form of all knowledge,
the categories are the cognitivist's version of the cosmos: they are
the ground of any possible knowledge, the outline of whatever is.
In the atmosphere of philosophical and economic individualism and
social alienation that followed the First World War, Durkheim and
his students sought to prove the necessity of sociality to human life,
and therefore of sociology as a discipline, by demonstrating that the
possibility of any general concept rests upon the prototype experience of collective force that primordially combines the many into the

one. In the early twentieth century, compelled by the vision of the
failure of civilization, Durkheim looked into the Aristotelian/Kant-
ian categories – those virtual forms, or empty moulds – and thought
he saw in them the shape of a tribal camp, the palpable image of a
coherent, functioning society. By contrast, in the late twentieth century United States, Lakoff and Johnson look into those same forms,
and in their illustrations we see some of the common, prosaic terms
of our world: the body, containers, getting into cars, getting jostled.
Compared with Durkheim and Mauss', most contemporary readers
would find Lakoff and Johnson's account more compelling. I suspect
the reasons include the greater compatibility of the latter with elements of regnant worldview, including our contemporary obsessions
with individualism, "personal space," and "body image."

The possibility for conflicting interpretations regarding prototypical experiences for fundamental categories inheres in the very
notion of fundamental categories. However Aristotle arrived at his
ten, the process certainly involved a deliberate search for those mental moulds of most general applicability. And here we return to the
basic problem: that to the extent that these fundamental categories are a successful winnowing of the most general properties of
things, and moulds of universal applicability, there is nothing they
are not congruent with. As congruency is one, if not the main, basis
on which these pairs of scholars argue for prototypicality, we are
left with no shortage of candidates for the experiential prototype
of the categories. To the well-established anthropological research
question of how different peoples impose their different categories
on the world, we should add the question of how different category-
theorists impose their different worlds on the category.

೦ಾ

Having focused on a comparison between the perspectives of Lakoff
and Johnson and Durkheim and Mauss, I would like now to comment on what I see as a kind of aberration, or perhaps transformation, that begins before but crescendos after *Philosophy in the Flesh*,
in a series of popular political writings in which Lakoff ends up less
at odds, than as convergent, with Durkheim and Mauss. Specifically, the role of family structure in the genesis of morality takes a
new form in Lakoff's political writings. In these he finds the difference between American conservative and liberal politics to lie in the

contrast between two different models of family structure, which he terms, respectively, "Strict Father morality" (conservative) and "Nurturant Parent morality" (liberal). Lakoff's notion that different political philosophies arise from different ideal images of family structure not only starts with the collective dimension of human existence, but, more specifically, it recalls Durkheim and Mauss' claim that different intellectual universes arise out of different social structures (or "social morphologies," as Durkheim and Mauss called them). It is as though kinship, having shrunk from the tribe to the immediate domestic sphere in the modern world, rebounds back to the tribe in the underlying, energizing metaphors of national political debate. The argument about contrasting family models is put forward by Lakoff in a series of political writings ranging from works that interdisciplinarily inclined academics or serious general readers might consult, to tracts that would have special appeal for grassroots organizers; these writings include *Moral Politics* (1996), *Don't Think of an Elephant!* (2004), *Whose Freedom?* (2006), *Thinking Points* (2006b), and *The Political Mind* (2008). It would be interesting to further explore the stylistic and strategic variations within these political works, but more to the point here are the qualities that unite them with one another and distinguish them as a group from earlier works by Lakoff and Johnson.

There is a shift in perspective as Lakoff moves into political pundit mode. With one important exception, which I discuss below, "the body" or "embodiment" as source of cognitive form and moral imperative – in other words, the linchpin of Lakoff and Johnson's earlier cognitive theorizing – recedes to a nebulous background, giving over its here-is-what-to-remember urgency to the two models of the family and their influence on politics. Why the shift? One possible reason has to do with the fact that Lakoff's intentions are now more applied than theoretical: the underpinnings may not be necessary for arguments aimed at the broadest possible political audience. Or it may be that the underpinnings are no longer the real underpinnings. I cannot help but wonder whether the body recedes because it is not the most fruitful source image for the political message that Lakoff seems to want to send: politics is about the relationship *between and among* people (between and among bodies, if you will), for which the image of a family provides a more potent generative bedrock – especially for the values of empathy and nurturance, which Lakoff puts at the top of the liberal (Nurturant Parent) political philosophy that he wishes to promote, and lower in the conservative (Strict Father) political philosophy, which he wishes to challenge.[21] This is speculative, of course, but no more so than the theories about category origins we have been considering. In part just because of the speculative character of their ventures, it is a stretch to see in either Durkheim and Mauss or in Lakoff and Johnson a scientific theory of the origin of categories. What we encounter is more on the order of origin myths, full of brilliant invention and experiential appeal, while infused with scientific rhetoric and the politics of the academy and the nation. The earliest recorded origin myths (the Babylonian Enuma Elish, for example, which served to legitimate the rule of the political paramount) as well as the earliest philosophical analyses of myths (Plato's *Republic* is the perennial example) are intensely political; there is no reason to suppose that contemporary forms of mythologizing should be any different.

The exception to the recession of the body in Lakoff's political writings lies in his analysis of a value that, in the earlier analyses considered above, does not figure among those ascribed to the body or to either the Strict Father or Nurturant Parent family, namely, the value of freedom. Lakoff's analysis of freedom is developed most fully in *Whose Freedom? The Battle over America's Most Important Idea* (2006), in which he claims that "the idea of freedom is felt viscerally, in our bodies, because it is fundamentally understood in terms of our bodily experiences" (2006:29). Following the procedure laid out in *Philosophy in the Flesh*, Lakoff begins with a discussion of the bodily experience of freedom, emphasizing especially freedom of motion, and then proceeds to derive from it philosophical inflections of the idea of freedom, emphasizing especially freedom of will and its framing within the tradition of faculty psychology: "In the Enlightenment, there was an elaborate metaphorical folk theory, called faculty psychology, in which the mind was a kind of society, with members who were individuals with different jobs. Among the members of the society of mind were Perception, Reason, Passion, Judgment, and Will. Perception gathered the sense data from the outside world; Reason figured out the consequences; Passion was the locus of desire; Will controlled action; and when Passion and Reason were in a standoff as to what Will should do, Judgment made the decision" (2006:33).

This account of faculty psychology brings the image of society into the philosophy of mind; in the following chapter we will further consider the "society of mind" metaphor as it has been refigured in recent artificial intelligence theory. What is important about the society of mind metaphor in Lakoff's political theorizing, however, is that it and the related metaphor of the "body politic" (2006:36ff.) open a direct route from politics to the body. This new route, based on the image of political society as a system of mutually sustaining parts analogized to the parts of a body or a mind,[22] bypasses the social collectivity with most immediate impact on the formative experiences of maturing individuals in American society, namely the family. This new route thus might seem to preclude, or at least evade, a possible Durhkeim-inspired alternate explanation of the experiential origin of the concept of freedom. Immediately following the sentence, quoted above, announcing the rootedness of freedom in bodily experience, Lakoff says this: "The language expressing the metaphorical ideas jumps out at you when you think of the opposite of freedom: 'in chains,' 'imprisoned,' 'enslaved,' 'trapped,' 'oppressed,' 'held down,' 'held back,' 'threatened,' 'fearful,' 'power-less.' We all had the experience as children of wanting to do some-thing and being held down or held back, so that we were not free to do what we wanted" (2006:29). Note that most of these opposites of freedom are terms whose first referent is human social relations, and that all of them have social relations as at least one possible ref-erent. The terms are accompanied by a "just-so" story, locating the birth of the idea of freedom in the child's experience of its opposite, i.e., constraint. The fact that Lakoff gives us a list of qualities denot-ing malignant, freedom-denying forms of social relations, opposing these to the body's visceral experience of freedom, stacks the argu-ment in favour of the body and against society.

But one could easily reverse all of this, as many thinkers have, and portray the body as a source of desires that, unmitigated, work against the ideal of the well-being of all. Society, by contrast, is the source, means, and guarantor of freedom: it offers the possibility of enacting protections against unscrupulous individuals, mechan-isms to adjudicate grievance, and, in language and accumulating cul-tural traditions, sources of intellectual and aesthetic stimulation and choice through which individuals make themselves into distinctive and effective persons. Lakoff's origin story about the child's first experience of constraint could be recast as the child's primordial experience of a protective nurturance that guarantees freedom from fear and the chance to mature. Much of this theme – that is, the freedom-enhancing side of protective social constraint – can actually be found in Lakoff's fuller analysis of freedom,[23] and it all fits rather well with the Nurturant Parent liberal political model that he advocates. Once again, there is an analytical choice of what to single out as experientially primary. But to go through Lakoff's arguments concerning freedom and counterpose a full Durkheimian alternative – that is, an account of the origins of freedom in irreducibly collective experience – would merely net another rehearsal, in a new register, of what has already been said above in the context of space, time, and causality.

୧୨

Some cosmological projections may in the end have little to do with cosmology in any direct sense. At the present state of the art, our most potent and encompassing projections may be those that take the form of a claim that certain experiences are the prototypical source of categorical structure in general and thus the framework of all knowledge. Our presently most potent form of projection, in other words, is perhaps not the experience we project onto all others, but rather the experience that we project as the experience that we project onto all others (for novelty's sake, I shall avoid labelling this as meta-projection). Rejection of the Kantian tendency to assume that the synthetic principles of cognition are inscrutable in favour of the view that they have a discoverable origin, results in these principles themselves, and even the generic structure of the category – such a prize these now are! – turning into a projective screen for the world as we know it, as well as our hopes, dreams, and desires for it – a screen as alluring as the starry sky above.

SURVIVING AND PREVAILING: THE WITTGENSTEINIAN REVOLUTION

In the foregoing analysis I have focused on issues connected with one of two main usages that theorists employ when they refer to Aristotelian categories: one usage designates a list of highly abstract, maximally applicable predicates or frames (time, space, cause, etc.) judged by Aristotle as intrinsic to the making of propositions; the other designates one of the specific frames that inheres in Aristotle's list: the notion of a class or set (or category in a narrower sense). In category theory these two usages often flow into one another.

By way of a complement to the foregoing, I offer in this final section a brief excursus into a third main usage of the term "Aristotelian" in relation to categories. A category is Aristotelian in this third sense when it is a category (or set or class) that admits no gradation of membership and thus operates binarily: any entity either is or is not a member of a particular category. This is the type of category used by Aristotle in his logic (and hence its designation); it corresponds to what Lakoff and Johnson in *Philosophy in the Flesh* call an "essence" prototype.[24]

In recent times a good deal of attention has been paid to sets that allow graded membership and "fuzzy" boundaries and that are thus different in principle from such binary or Aristotelian categories. Particularly influential in promoting the recognition and importance to human cognition of non-Aristotelian categories has been the work of Eleanor Rosch (1978), though founder's credit is often accorded to the philosopher Ludwig Wittgenstein, who wrote:

> Consider for example the proceedings that we call "games". I mean board-games, card-games, ball-games, Olympic games, and so on. What is common to them all? – Don't say: "There *must* be something common, or they would not be called 'games'" – but *look and see* whether there is anything common to all. – For if you look at them you will not see something that is common to *all*, but similarities, relationships, and a whole series of them at that ...
>
>   I can think of no better expression to characterize these similarities than "family resemblances"; for the various resemblances between members of a family: build, features, colour of eyes, gait, temperament, etc. etc. overlap and criss-cross in the same way. – And I shall say: 'games' form a family. (1968:31–2)

In an earlier work, Lakoff (1987), taking his cue from Wittgenstein and championing the work of Rosch, has portrayed the contrast between the category based on "family resemblances" and the clearly bound Aristotelian category as the basis for a major revolution in cognitive science.[25] But I am intrigued by the substrate that survives this purported revolution, for once again both antipodes implicate the image of a family. Aristotle's works on logic and classification are laden with the terminology of kinship (starting with "*genos*" or kin/family, rendered in Latin and English as "genus"), yet when Wittgenstein offers an alternative to Aristotle, the alternative, too, is portrayed through an image of kinship, i.e., family resemblance.

Another dimension of the appeal to kinship in the image of family resemblance confirms a continuing relationship between *kinship* and *categories* in recent times. Specifically, theoretical developments within the comparative study of kinship systems – a specialization that declines and then recrudesces within social and cultural anthropology – have undergone some theoretical turns that parallel the broadening of theories of categorization inspired by Wittgenstein and Rosch. The boom in comparative kinship studies in the midtwentieth century was given impetus by the idea promoted by Durkheim that kinship systems are for societies what skeletal structures are for living organisms; as organisms are scientifically classified and compared through skeletal types, societies can be classified and compared cross-culturally through kinship-system types. A.R. Radcliffe-Brown (1965) adopted this idea and developed an influential set of methods and theories about kinship, for the most part assuming that kinship systems would be unilineal (that is, based on tracing descent dominantly through *either* the male line *or* the female line). Partly as the result of the arguments of Polynesianists, notably Raymond Firth (1963; cf. Schwimmer 1978 for Maori), kinship studies more recently have accepted the possibility that the principles on which kin-groups are based may not be as neatly unilineal as once assumed: individuals are often found to claim filiation of great genealogical depth from both parents (sometimes compounding at ascending generations), making clan membership fuzzy, negotiable, and shifting. This realization appeared just as cognitivists were recognizing that folk categories in general often allow inclusion through a plurality of routes rather than by binary criteria. Thus in these two realms – cognitive theories of folk categories, and anthropological theories of kinship – the findings converged: humans create classes characterized by flexibility, gradation of membership, and "fuzzy boundaries." It seems that as goes theory of kinship, so goes theory of the category – or vice versa.

There are some interesting moments in these kinship debates. One of these concerns anthropologist Claude Lévi-Strauss' discussion of the "house" in the context of kinship; house makes reference to a seeming kin-group constituted through several inconsistent modes of filiation. As noted above, under the formative influence of Durkheim's work on kinship, elaborated by Radcliffe-Brown, the anthropological study of kinship had come to be dominated by the ideal of the clearly bounded Aristotelian category, or in other words, the assumption that all kinship groupings necessarily follow binary Aristotelian logic (any individual is a member of *either* this clan *or* that). Considering (as discussed above) that Durkheim wanted to show that Aristotelian logic derived from the forms and practices of social life – for him epitomized in the tribal genealogy – one can see why: if Aristotelian logic is to be traced to tribal kinship practice, then tribal kinship had better be Aristotelian in principle![26]

As noted above, anthropologist Lévi-Strauss caused a flurry – within the rarefied milieux of kinship theory – by invoking as a kinship term a concept that relates to kinship only metonymically – the term "house" – in order to designate a type of social grouping, encountered in various parts of the world, which is based on several different modes and degrees of filiation rather than a unilineal principle of descent. It is surely an indication of the degree to which anthropological studies of kinship had been skewed in an Aristotelian/binary direction that Lévi-Strauss resorted to a term with no intrinsic connection to kinship (i.e., house) in order to capture the possibility of fuzziness that to the naive outsider – the philosopher Wittgenstein – seems to have been the very soul of kinship. Why else would the latter have adopted family resemblance as the emblem of classificatory fuzziness?[27]

Another quirk can be found in anthropologist David Schneider's colourful and radical critiques of the anthropological study of kinship (1968; 1984). Schneider was so struck by the variability of so-called kinship systems (and by theoretically anomalous points, some of them quite charming, such as the insistence of some informants that the dog was a member of the family) that he was led to question the very existence of this particular category of ethnographic phenomena – wondering whether kinship was an ethnographer's illusion. But it is important to note, once again, that this skepticism developed in the context of an anthropology predisposed to search for an Aristotelian category: an Aristotelian essence of kinship. One wonders whether it would have made a difference if Schneider had, instead, approached the variability of kinship systems in terms of the possibility of discovering areas of graded overlap in the manner of Wittgenstein's family resemblances. That would have been an interesting situation: kinship salvaging itself as a topic of theoretical discourse by serving as its own naive metaphor. This position has much to recommend it; and it is partly in view of this possibility that, in the face of Schneiderian skepticism, I still dare to think that kinship is a genuine topic of ethnographic exploration (even while, admittedly, remaining fuzzy about what I would expect to encounter within it). Schneider was no mythologist, yet other thinkers noticed the implications for mythology of his radical critique of kinship theory. Judith Butler (2000), for example, enlists Schneider's critique in an influential and thought-provoking re-evaluation of the sagas of Oedipus and Antigone around their confused kinship.

Though in contemporary Western society our most widely shared public cosmologies tend not to be heavily reliant on kinship images (with exceptions, of course: contemporary scientific cosmology is replete with kinship metaphors such as "baby universes," "still-born universes," and so on), we nevertheless have maintained, even in our most rationalistically inclined intellectual pursuits, numerous forms of speculation to the effect that the cosmos, or its building block, the category, and kinship somehow go together; like Zeus, we are not able to let go of our kinship web. These speculations range from the claim that the cosmos or the category is a projection or formalization that abstracts the structure of kinship, to the recurrent, sometimes almost subliminal, expectation that kinship will serve as a reliable metaphor for the creation of order in general. From this point of view alone it should come as no surprise that Lakoff, in the most popularizing of his works, carries his theoretical speculations about the genesis of moral reasoning in kinship into the more specific claim that the political polarity in contemporary American politics is best grasped through contrasting models of familial order (the aforementioned Strict Father and Nurturant Parent models). As noted above, Durkheim and Mauss suggest that, formal structure aside, *our very predisposition* to approach the cosmos as though its entities are related in a web of mutual influence, rather than as discrete isolates, flows from the experience of being born into a human situation with this character.[28] Finally, it is almost as though Kant anticipated Durkheim's sociologizing move, on this point, by naming one of his fourteen a priori categories – specifically, reciprocal-causality – as "community" ("*Gemeinschaft*"), thus pre-positioning a moral concept for projective extension to the physical world.

It is difficult to imagine any method that could definitively move any such ideas beyond the realm of conjecture or projection. But, given the constancy, beginning even with Copernicus (who led us out of our anthropocentrism), with which our intellectual abstractions have been attended by images of kinship, it would be equally difficult to find (should we want to) a method that would cause the latter images to definitively depart – a method that would, in other words, fully put to rest our Zeusian affliction. The borderline of the great Copernican divide, the point at which we learned to think of the universe scientifically rather than mythically – as a mechanism rather than as the rest of our family – is, logically and temporally, a fuzzy one. Fuzziness as an intellectual mood has, as one of its grammatical manifestations, double-negative constructions, for it is somehow less committal to say that "X is not without" than to say that "X *is*". I therefore find it not inappropriate to conclude with a choice Lacanian triple-negative: "Not that what isn't Copernican is absolutely unambiguous" (1991:7) – which I take to be a French critical theory way of saying that what we mean by Copernican may not be entirely clear.

5

Descartes Descending or the Last Homunculus: Introversive Anthropomorphizing in Popular Science Writing

Thus instead of being peopled with a noisy bustling crowd of full-blooded and picturesque deities, clothed in the graceful form and animated with the warm passions of humanity, the universe outside the narrow circle of our consciousness is now conceived as absolutely silent, colourless, and deserted ... Outside of ourselves there stretches away on every side an infinitude of space without sound, without light, without colour, a solitude traversed only in every direction by an inconceivably complex web of silent and impersonal forces. That, if I understand it aright, is the general conception of the world which modern science has substituted for polytheism.

> Sir J.G. Frazer, *Man, God and Immortality: Thoughts on Human Progress*

Long ago when the members of the human body did not, as now they do, agree together, but had each its own thoughts and the words to express them in, the other parts resented the fact that they should have the worry and trouble of providing everything for the belly, which remained idle, surrounded by its ministers, with nothing to do but enjoy the pleasant things they gave it. So the discontented members plotted together that the hand should carry no food to the mouth, that the mouth should take nothing that was offered it, and that the teeth should accept nothing to chew. But alas! while they sought in their resentment to subdue the belly by starvation, they themselves and the whole body wasted away to nothing. By this it was apparent that the belly, too, has no mean service to perform: it receives food, indeed; but it also nourishes in its turn the other members, giving back to all parts of the body, through all its veins, the blood it has made by the process of digestion; and upon this blood our life and our health depend.

Aesopian fable of "The Belly and the Members"[1]

Even if, as diagnosed by Frazer, the world outside "the narrow circle of our consciousness" is to be regarded, in the era of science, as "deserted" rather than bustling with the "warm passions of humanity," may we at least remain confident that matters inside the circle will escape the same fate? The answer is not clear. In this chapter I address one element from the realm of artificial intelligence (or AI) research: the theme of homunculism. I am drawn to the homunculist theme in part through the obvious resonances with mythological traditions of "little men" (fairies, elves, and dwarves) and by folk traditions involving anthropomorphizing of body parts (most famously in the fable of the "belly and members"), but even more so by a more abstract consideration. Specifically, I am intrigued by the parallels between the idea of a "homunculist fallacy" as it is sometimes called (in AI research and cognitive theory more generally),[2] on one hand, and, on the other, a long-established theme in the theory of myth, a theme that might be summed up as the "anthropomorphic fallacy."

ARTIFICIAL INTELLIGENCE AND THE HOMUNCULIST FALLACY

There seems to be general agreement about what the homunculist fallacy consists of, namely, *purporting to explain* or account for intelligent behaviour merely by referring it to inner beings or agencies that possess the capacity that is to be explained – an inner little man that intelligently reads inputs and decides on intelligent courses of action. The philosopher Gilbert Ryle is often credited with calling attention to the problem of infinite regress inherent in appeals to such inner agents and voices. In addition to the charge that it really explains nothing, the fact that homunculist imagery is frequently invoked in everyday discourse also opens up such imagery to the charge of being mere "folk psychology."

Fully cognizant of these criticisms, some artificial intelligence researchers nevertheless regard homunculism as holding out a promising research strategy. One part of the strategy involves attempting to surmount the problem of regress through a sort of division of labour. The idea is that it is possible to escape regress – or at least makes "progress" in explaining (as opposed to merely transferring) intelligence – by analytically dividing up intelligent tasks, positing teams of homunculi that are progressively less knowing at each level of analysis, with the goal that the final, most limited homunculi be implementable by machine or a neural structure. Daniel Dennett, a well-known advocate of the homunculist approach, offers an exemplary portrait of the strategy in the following passage:

> Homunculus talk is ubiquitous in AI, and almost always illuminating. AI homunculi talk to each other, wrest control from each other, volunteer, sub-contract, supervise, and even kill. There seems no better way of describing what is going on. Homunculi are *bogeymen* only if they duplicate *entire* the talents they are rung in to explain ... If one can get a team or committee of *relatively* ignorant, narrow-minded, blind homunculi to produce the intelligent behaviour of the whole, this is progress. A flow chart is typically the organizational chart of a committee of homunculi (investigators, librarians, accountants, executives); each box specifies a homunculus by prescribing a function *without saying how it is to be accomplished* (one says, in effect: put a little man in there to do the job) ... Eventually this nesting of boxes within boxes lands you with homunculi so stupid (all they have to do is remember whether to say yes or no when asked) that they can be, as one says, "replaced by a machine". One *discharges* fancy homunculi from one's scheme by organizing armies of such idiots to do the work. (1978:123–4)

In AI literature one comes across many variations on Dennett's statement; particularly recurrent are the themes of making progress in accounting for intelligence through decomposing functions into simpler sub-functions, and termination of these sub-functions in a neural or machine-emulatable bedrock. I will return to Dennett's statement in the course of my analysis.

THE JOURNEY TO NEURONOPOLIS: A CINEMATIC DIGRESSION

My roundabout journey to the topic of AI homunculism began with a cultural "descent" of my own, from rarified academic discourse to popular science education films. Since the films in question provided the ultimate inspiration for the ideas offered in this chapter, a brief digression about them will help to frame the presentation that follows.

As part of a larger interest in processes of anthropomorphism, I recently took a trip down memory lane to revisit a series of films produced in the 1960s under the sponsorship of what was then Bell Telephone; although silly by today's standards, these films had enormous impact in stimulating the growth of popular science through visual media – they are modern classics. My own favourites from this series are *Our Mr Sun*, which deals with the solar system and physical cosmos, and *Hemo the Magnificent*, which deals with the blood, heart, and circulatory system. They were aired on television during family viewing hours; and since they became a fixture in Bell's public relations program, I and others of my generation will remember seeing them once again in the grade school assembly hall – the movie projector clicking (and sometimes smoking) in the background.

Key to both of these films is a dialogue between scientifically informed, actual humans, on one hand, and, on the other hand, anthropomorphized versions of various objects from the natural world that the humans are investigating – the latter brought to life through old-fashioned (i.e., pre-computerized) studio animation. The anthropomorphizing of "matter" is a fascinating business: in these films new scientific findings are presented in explicit opposition to older mythological notions – notions discredited by science – even as the natural phenomena under investigation are re-mythologized in the new medium of cartoon animation.[3]

Our Mr Sun, for example, is built around the plot of a scientist and a filmmaker (both played by actual humans) collabourating to make a movie about the sun. The filmmaker proposes to the scientist the gimmick of an animated Mr Sun. However, his cartoon mock-up of Mr Sun – a sort of smiley face with rays – spontaneously comes to life, adopts the trappings of a movie director (including a director's folding chair and "sunglasses," so to speak), and declares that he himself should direct his own movie. Resentful of the ways in which he is portrayed in science, Mr Sun proposes a screenplay (at this point we thus have a movie within a movie) that will feature not the findings of science but, rather, representations of the sun drawn from mythico-religious traditions. In one scene a sequence of cartoon-animation mythological images (including an Eskimo figure of the sun as a "little man wearing thick furs" and the Greek figure of "Apollo racing across the sky in his fiery chariot") pass in review. Mr Sun disapproves of the cold, analytical way he is portrayed in science, but approves of these mythico-religious portrayals – although admitting that some of them contain "pretty odd notions" about him. One tends not to see one's own mythology: within the parade of images pronounced as odd by Mr Sun, one does not see *Mr Sun the movie director*. In the course of the film Mr Sun is gradually won over by the scientist and filmmaker to the cause of science.

Here I will focus on one point that is brought home in moving from the film *Our Mr Sun* to another film in the same series which deals with blood and circulation, *Hemo the Magnificent*. Specifically, moving from one to the other reveals a common "formula" uniting the two films, *even though* one dealt with the cosmos outside of us and the other with our internal organs. On one level, the common formula is related to the fact that both were directed by Frank Capra, a director with a reputation for wringing the most out of a successful formula.[4] But beyond the director, I was struck by the apparent ease with which techniques for anthropomorphizing things outside of us, such as the sun and other stars and planets, could be redeployed to portray our innards, specifically the various functions of the brain and other internal organs.

Consider a knockout scenario from *Hemo the Magnificent* that attempts to explain the role of the circulatory system in maintaining consciousness. This scenario portrays, again through studio animation, what happens when a prizefighter takes a direct blow to his head and loses consciousness. The scenario is based on the structure and functioning of a modern business corporation. A little man in a business suit (Mr Big) sits at a desk inside the brain of a prizefighter. Mr Big monitors inputs and sends messages to other little men distributed throughout the prizefighter's body, including a set of men dressed in gym clothes – minuscule myocardial musclemen – who push squeeze-bulbs (reminiscent of those of the sphygmomanometer, the device typically used to measure blood pressure) to move the blood. When the prizefighter takes a hit, the brain "dispatcher," a 60s-style switchboard operator, topples and passes out and, due to the disrupted communications, the capillary gatekeepers, little blue-collar men who operate valve levers, relax and allow too much blood to flow to the fighter's abdomen. Too little blood then reaches the brain, so Mr Big passes out and topples over inside the office in the brain. Finally, emulating what has already happened in miniature on the inside – first to the dispatcher and then to Mr Big – the prizefighter loses consciousness and topples over.

Mr Sun, as movie director, barks orders on the cosmic level; Mr Big in the brain barks orders on the level of a generic human body. Beyond this, the two scenarios have, on the surface, little in common. Yet the viewing experience is that of seeing two versions of one film. The core feature that deeply relates the two films is human society as a source of imagery for portraying the processes of the natural world. The common framework of human interaction allowed Capra to flesh out the two scenarios using a single bag of tricks, which film critics have come to refer to as "Capra-corn," or in other words, the range of maudlin but powerful human situations that Capra taps to move viewers emotionally. Weaving through many of these situations is the powerful and intricate reality of social hierarchy – a reality tapped as much by Capra as by archaic mythologies to portray the order and working of the natural world (recall the discussion of kinship in the previous chapter).

It was the experience of a common formula in these two primitive popular science films, deployed in one case to the outer cosmos, in the other to the world within our bodily borders, that inspired the present chapter. Vaguely familiar with a technical and serious "homunculism" theme in cognitive theory, and as if egged on by the cartoons, I decided to have a look at anthropomorphic introversion – or as pioneer AI researcher Marvin Minsky enticingly terms it, "the myth that *we're* inside ourselves" (1986:40) – considering inward projections of human images vis-à-vis the outward projections more familiar in the field of mythology as traditionally defined.

Introversive anthropomorphizing is by no means limited to AI homunculism. Growing knowledge of physiology opens new possibilities for our anthropomorphizing inclinations: genes, memes, neurons or other cells, and quarks or strings as components of body/mind, among other "parts," have proven to be appealing targets. Recently a debate erupted around the claim that scientific observations of reproductive fertilization were tainted by a tendency to project gender stereotypes of males as active and females as passive into microscopic observations of the interaction of sperm and ova (for example, see Martin 1996; Gross 1998). There are interesting borderline cases, such as Richard Hofstadter (*I Am a Strange Loop*

[2007]) talking about the strange loop that *is* a given individual as though it is also *in* that individual's brain along with the strange loops of the individual's significant others. Popular culture offers an array of variations on little people inside (for starters, the cult films *Double Indemnity, Fantastic Voyage, Being John Malkovich, Inception*). There is a longer history and wider anthropology to be investigated as well. The first-century BCE poet Lucretius argued that thinking, along with all other processes found in nature, occurred through the interaction of material atoms; Duncan Kennedy (2008) calls attention to Lucretius' anthropomorphizing of such atoms and offers some comparisons to the contemporary sperm and ova and selfish gene debates. Perhaps the inclination to anthropomorphize at the micro-level has something to do with, or carries further, the broadly documented human inclination to bring amorphous mental or emotional phenomena to a focus at the level of a specific internal part, such as the heart or liver.

The general thesis of this chapter is that introversive anthropomorphizing involves imperatives and constraints that both parallel and significantly differ from those involved in the outwardly directed anthropomorphizing common to mythological portrayals of the world outside us. Although in this chapter I analyze only one arena of introversive anthropomorphizing, the specialized if not rarefied realm of AI homunculism, my hope is that the approach offered – built around a close juxtaposing and comparison of introversive and extroversive anthropomorphizing – will contribute toward a method and perspective for analyzing a broader trend in popular science writing, one that is likely to grow as works about our "insides" take on increasing prominence in this genre.

One reason that AI homunculism can admirably serve as paradigmatic case is that its practitioners have already developed a useful self-critique in response to the aforementioned homunculist fallacy or the charge of folk psychology; and their critique resonates in interesting ways with the critique that has developed over centuries and across several academic fields regarding anthropomorphizing of the world outside of us (in which traditional mythologies are often invoked as exemplar). Introversive anthropomorphizing is not necessarily limited to works of popularization, but popularizing ambitions clearly give this process an enormous boost in urgency and rationale.

ANTHROPOMORPHISM: EXTROVERSIVE AND
INTROVERSIVE

My purpose in this chapter differs somewhat from that of the previous chapters, for in those I focused on bringing to the surface anthropocentric strategies that are largely unacknowledged by their authors. By contrast, the anthropocentric (or specifically anthropomorphic) character of the homunculist strategy of artificial intelligence research is openly acknowledged, often accompanied by the sentiment that the homunculus is a mere stop-gap heuristic necessitated by the rudimentary state of brain science, or by other disclaimers. My focus in this analysis is not the anthropocentric or anthropomorphic character of homunculist analysis, but rather the relationship between the inward-directed or introversive anthropomorphism of homunculism, on one hand, and, on the other hand, the outer-directed or extroversive anthropomorphism characteristic of the visions of the cosmos typically referred to as mythology. On the whole, AI homunculists demonstrate no particular expertise in theory of myth, and I am not aware of any previous mythologist who has analyzed AI homunculism; yet these fields have much to learn from one another. In an epilogue to this chapter I ask why these two topics have remained unconnected, passing each other as ships in the night.

Juxtaposing introversive and extroversive anthropomorphism reveals a number of parallels and intriguing twists in the process of anthropomorphizing and in the intellectual traditions that have grown up around the analysis of these processes. My goal in this chapter is to direct this juxtaposition toward the broadening of our understanding of anthropomorphizing processes.

For purposes of this analysis, the following definitions (consistent with usages in previous chapters, though more finely drawn here) are employed:

- "Anthropomorphism" is the encompassing term, referring to the projection of human forms and characteristics into non-human realms, whether extroversively or introversively.
- "Mythology" refers to extroversive anthropomorphism.
- "Anthropomorphic fallacy" refers to the alleged fallacy of extroversive anthropomorphism (or the alleged fallacy of mythology).
- "Homunculism" refers to introversive anthropomorphism.

- "Homunculist fallacy" refers to the alleged fallacy of introversive anthropomorphism.

There is one obviously inelegant point in the above definitions, specifically, that anthropomorphism as a process encompasses both introversive and extroversive versions, while anthropomorphic fallacy is limited to only the extroversive fallacy (i.e., mythology). After experimenting with several definitional schemes, I have concluded that such awkwardness is unavoidable. The notion of anthropomorphically induced error has grown up in close association with the study of mythology, which in turn has been thought of, for the most part at least, as outwardly projected anthropomorphism; due in part to this historical association, anthropomorphic fallacy is a more appropriate designation than mythological fallacy, a term with no historical resonance. In the end the terminology is arbitrary, but the focus of my study is not: I compare introversive with extroversive deployments of the seemingly unified human impulse to portray natural processes on everyday models of ourselves. As noted earlier, following the comparisons, I offer in an epilogue some thoughts about why the respective thinkers concerned with these two deployments have had so little to say to one another.

ℭ

At the risk of being needlessly cautious, it should be emphasized at the outset that none of the AI homunculists are looking for actual little physical men in our brains. Positing a homunculus amounts to positing a mechanism or "unit" (see Dennett 1991:261) with a function whose precise workings may not be entirely understood. AI homunculi are often said to be metaphorical; I accept this claim although I think that in some invocations AI homunculi also carry additional connotations beyond those typically implied by metaphor; these cases will be considered presently. Yet let us be clear that no one is positing actual little men in our heads. Do AI homunculi then have *anything* to do with the animated gang of Hemo through which I was led to this topic? Yes! For reasons of audience and context, AI homunculism has reasons to distance and minimize its imagery while Frank Capra had reason to ham it up; but in both cases we are clearly dealing with the same metaphors and at least substantial overlap in the reasons for invoking them. That the animated scenarios of little men inside us (in Capra's *Hemo*) are explicitly intended as metaphors is indicated through periodic juxtaposition of the animated circulatory system with photographs of the *real* circulatory organs (filmed during open-heart surgery – a rather frightening scene for grade-schoolers!). Of all who have been implicated thus far – AI homunculists, Frank Capra, I and my classmates in grade school – nobody is claiming that there are actual, literal little men inside our heads. We all get the idea of a metaphor of mental control-functions, whether in its Hollywood or academic AI form.

But if contemporary homunculism (in both its AI and Hollywood incarnations) is predicated on a metaphorical intent, what can be said about myth in this respect? A mythologist not infrequently encounters formulations in which myth and metaphor are portrayed as conceptual opposites: myths, it is alleged, are based on literal belief, whereas metaphors have a figurative, heuristic, character – and the latter may add a bit of fun to a serious project. Indeed, one of the reasons for the lack of any signs of mutual interest between AI homunculism, on one hand, and theory of myth, on the other, might be a reluctance by AI homunculists to admit that the knowing, strategic anthropomorphism of homunculism could have anything to do with the allegedly literal, innocent anthropomorphism of pre-scientific mythologies. At a later point I raise the question of whether it is appropriate to assume such a sharp dichotomy between myth and metaphor on this issue. But at the outset I want to say that *even if* myth and metaphor diverge categorically on a literal vs. figurative axis, they may *still* have much in common, particularly as regards the choice and structuring of imagery. A scientific AI homunculist, when working out a plausible model of the brain's structuring of a certain intelligent task, seems no less willing to be inspired by familiar, everyday hierarchical structures (say, of boss or supervisor and various categories of underlings) than is the proverbial pre-scientific, mythopoeic mind asking why the stars move – even though for the first, the exercise is a thinly engaged-in fiction surrounded by disclaimers, while for the second, it quickly hardens, at least according to a common stereotype about mythology, into a literal belief that its creators turn around and worship. I suggest that the efficacy of anthropomorphic imagery in bringing coherence to the world may in the end have little to do with whether such imagery is subscribed to as literal belief or heuristic metaphor.

The objection might be raised that the problem of literalism vs. metaphor could have been avoided by leaving AI out of my analysis entirely, and comparing, instead, instances of introversive and extroversive anthropomorphism that have both been drawn from pre-scientific traditions. My response is that while one does encounter in folk psychology instances of inner "voices," "demons," divided "selves," and other such notions, I have as yet come across nothing in folk psychology as *systematically* parallel to mythology (in its usual sense) as are the portrayals of the mind that have been created by the scientific psychology of AI research. Whatever be the fate and final significance of homunculism within AI research, there will always remain an intriguing point for the history of mythology. That is, quite possibly *the* most enthusiastic and colourful elaboration of internal natural processes on the image of the doings of human-like beings and their society, an elaboration that, at the least, humbles the traditional fable of the "belly and members," is to be found not in archaic mythology or folk wisdom but in an enterprise defined in opposition to these.

MYTHOLOGY AND THE ANTHROPOMORPHIC FALLACY

The idea of anthropomorphism is so closely associated with the idea of mythology that in many contexts the two terms are nearly synonymous; the familiar notion of myth supposes that it is a highly humanized, personalized account of the origin and workings of the cosmos and nature. For many readers, ingenious anthropomorphism is one source of myth's delight. However, the history of theory of myth also contains a long-standing criticism of myth that takes aim at precisely this anthropomorphic inclination. Two basic versions of the anthropomorphic fallacy have been articulated, one associated more with theology, the other more with science;[5] I will summarize these in turn.

The *locus classicus* of the first, which I will refer to as the theological version of the anthropomorphic fallacy, is the writing of ancient Greek philosophers, most notably Xenophanes and Plato, though the tradition has a robust life in much of subsequent Western theology. The fallacy, in this view, amounts to the mistaken notion that it is possible to portray the gods or God on a human model. For the gods, argue Xenophanes, Plato, and their followers, are in every way superior to humans: more encompassing, stable, and rational, and less given to appetite.

The *locus classicus* of the second, which I will call the scientific version of the anthropomorphic fallacy, is the eighteenth-century Enlightenment vision of the progress of science and the demise of traditional superstition. Frequently implicated in this view is the notion that mythology recedes when the idea of a universe of personalized forces gives way to the idea of a predictable mechanistic causation, a vision epitomized in Newtonian planetary mechanics.

The concern with the anthropomorphic fallacy that operates in the latter, or scientific, context differs in two main ways from the parallel concern that operates in the theological context. First, while the theological concern with anthropomorphism is directed toward anthropomorphic distortions of deities, the scientific critique is mainly concerned with anthropomorphic distortions of the physical world and its workings, including those distortions allegedly introduced by the very idea of deities. From the first perspectives, deities are beings that humans distort anthropomorphically; from the second, they may be seen as figments of imagination introduced to the cosmos by our anthropomorphizing inclinations. Second, in its scientific version the anthropomorphic error is often conceived of as carrying a second blow: we learn that the self-image we have unwarrantedly projected onto other things in the cosmos is – additionally – an unwarranted image of ourselves. That is, our favourite self-images are laden with phantom notions of spiritual substance and properties – conceits such as free will – that are mistaken not just in their extended applications but at their source: what we have been mistakenly projecting onto the rest of the cosmos is not just ourselves but mystified, overly exalted notions about ourselves.

A consummate expression of the anthropomorphic fallacy in its scientific version is found in the writings of nineteenth-century social evolutionist E.B. Tylor, a figure whose views about the relationship of science and myth still exert considerable influence.[6] For Tylor, the human inclination to animism and anthropomorphism turns out to be an error resting on another, larger error. The larger error is that of ontological dualism (or Cartesian dualism), the conviction that the universe contains two irreducible substances, matter and spirit/mind substance. Dualism reflects a human inclination to suppose ourselves possessed of higher powers that bring control over the world of matter, even as we project these powers onto non-human things to gain access to and familiarity with them under the supposition that we are their adequate model. To combat such tendencies, Tylor preached a kind of reverse projectionism: instead of foisting our self-conceit on nature, we should learn to conceptualize the workings of our minds on the image of the regular motion of the planets and the waves of the ocean, emulating the successes that a monistic, materialist science has brought to our understanding of things outside of us.

൦‍൭

In what follows, I offer the observations of a mythologist regarding homunculist issues, especially the homunculist fallacy, in AI and cognitive theory. Rather schematically I suggest the ways in which the idea of such a fallacy recapitulates terrain familiar to a mythologist amidst obvious lines of divergence. First of all, there are some direct parallels between these two regions, critical theory of myth and AI homunculism. But then there are other similarities between these two regions that take the form of *inversions*, a term that I employ broadly to designate analytical situations in which, in passing from one side of a conceptual divide to the other, one encounters the same problem in some sort of mirror image.[7] Here, the analytical divide that occasions the inversions is the divide between the outer vs. inner direction of our anthropomorphic gaze, our epidermis marking the approximate line of demarcation. In yet other instances, the relationship between mythology and homunculism is interestingly asymmetrical. What follows are discussions of the relationship between issues in mythology and/or theory of myth, on one hand, and issues in homunculist AI, on the other, grouped into parallels, inversions, and asymmetries. Comparison illuminates the underlying strategy that unites the enterprise on both sides, that of invoking familiar human imagery to make some part of the natural world comprehensible. To follow my arguments it is essential that the reader keep in mind the definitions of key terms that I offered above, since I will use key terms (such as "mythology," "anthropomorphism," and "anthropomorphic fallacy") in ways that are more technical and narrow than common usage.

FIRST PARALLEL: THE SIMILAR SEMANTICS OF
"ANTHROPOMORPHIC FALLACY" AND "HOMUNCULIST
FALLACY"

The errors called to attention in the anthropomorphic fallacy and the homunculist fallacy are largely the same error. In broad terms, both fallacies amount to the charge of applying a human model in cases where this model fails to provide a correct description or explanation of the phenomenon in question. The two fallacies also share an interesting quirk: in both cases, the error is designated by a term whose immediate referent is physical form; but, upon close inspection, in *neither* case is physical form what the alleged fallacy is really about. Stewart Guthrie's *Faces in the Clouds* is an extensive work aimed at exploring the process of anthropomorphizing and surveying instances of it (with a few exceptions, which are considered below, his examples are instances of what I have called extroversive anthropomorphism). Guthrie's opening line contains a terse definition of anthropomorphism, i.e., "the attribution of human characteristics to nonhuman things or events" (1993:3). However, although the term "anthropomorphism" is built around the root "morphology" (with its connotations of physical form) and although Guthrie's survey contains a number of photographs of non-human objects such as implements and containers shaped to physically look human in some way, Guthrie's own theoretical discussion is not concerned with human physical morphology. Instead, Guthrie repeatedly mentions social structure, ability to form social relationships (especially of exchange), personality, language, and symbols use. The essence of anthropomorphism, the core of its error, for Guthrie and many others, turns out to be the transference of rather high-level human mental capacities to non-human nature; transference of human *form* is important mainly as an index of these mental capacities.

The same can be said for the homunculus image in the sphere of contemporary AI research. Through its longer history, the term "homunculus" has designated a physical form, a diminutive "little man" that figures in the reproductive genesis of physical form. Yet, as noted above, in the context of contemporary AI research no one is interested in actual physical little men. The homunculus is an emblem or metaphor of high-level mental qualities (or parts of them) not unlike those of concern to Guthrie. Complex, intelligent behaviours, including capacities for language and other symbolic behaviour, are the real concern; homunculi merely index these.

Differing from Guthrie, AI researchers sometimes glide over the details of social structure and exchange among the homunculi they invoke in attempting to account for specific intelligent processes – a lacuna that may be fateful. That Guthrie puts social organization, especially exchange relationships, on the same level as more obviously symbolic processes (such as language use), would seem to reflect Guthrie's anthropological background, for in recent decades many socio-cultural anthropologists, recognizing semiotic processes embedded in social structure and exchange relationships, have tended to treat these as language-like systems, sometimes even directly applying to them the theoretical apparatus of formal linguistics. But, given that even sophisticated patterns of social structure and exchange are, like language, experienced by humans through rules that are unconscious and/or tacit, the social structure of AI homunculi offers one of the more obvious routes through which properties of intelligence – the purported *explanandum* of AI research – might instead be subliminally slipped into the *explanans*. I will return to this theme.

SECOND PARALLEL: TRADITIONAL MYTHOLOGY AND AI HOMUNCULISM BOTH DRAW HEAVILY ON THE IMAGE OF SOCIETY

The realms of imagery drawn on in AI homunculism are also very familiar realms in mythology. Mythology, in portraying the workings of the cosmos, and homunculism, in portraying the workings of the mind, appear to be primarily indebted to the same image: human society, with emphasis on social hierarchy and division of labour as organizing principles. In homunculism, as in many mythologies, we encounter elaborate human-like hierarchies organized to explain the working of a whole and its parts. Many mythico-religious pantheons are organized as if to describe and explain the workings of the natural world and/or the cosmos. Such pantheons tend to be structured with more powerful and more knowing gods at the apex, and more specialized functionaries toward the bottom. The hierarchy is typically portrayed through kinship relations, often on the principle that the oldest (and therefore genealogically deepest and most encompassing) gods are the most knowing and powerful. Such principles are familiar in Greek (e.g., Hesiod) and many other mythologies (including the Maori examples discussed in chapter 4).

While functionally organized hierarchies of mythico-religious pantheons often take the form of kinship structures (with systems of authority and functional roles allocated according to genealogical position), the functional hierarchies of AI homunculism, by contrast, are typically portrayed on the model of modern bureaucratic organization, including flow charts that diagram the interactions of CEOs, librarians, middlemen, facilitators, and so on.[8] It is as if AI analysts compete to demonstrate cleverness in applying the imagery of contemporary bureaucracy to mental processes. The following is one of many possible examples, in this case having to do with face recognition. This example intermixes terms that connote a human operative ("public relations officer"), terms that are ambiguously human ("analyzer"), and terms that are bio-mechanistic ("motor subroutines"), as if to suggest the ultimate transposability of such terms:

> The executive routine will direct a *viewpoint locator* to look over the perceptual display, and the viewpoint locator will sort the input into one of the three possible orientation categories. The display will then be shown to the appropriate *analyzer*, which will produce as output a coding of the display's content. A *librarian* will check this coded formula against the stock of similarly coded visual reports already stored in the organism's memory; if it finds a match, it will look at the identification tag attached to the matching code formula and show the tag to the organism's *public relations officer*, who will give phonological instructions to the *motor subroutines* that will result in the organism's publicly and loudly pronouncing a name. (Lycan 1987:43–4)

At moments AI homunculism seems to be sociology all over again, reinvented at the level of the organization of the individual generic mind.[9]

As noted above, mythology tends to be associated with images of kinship, and AI homunculism with images of modern bureaucracy and with contemporary social idioms such as corporate-speak. But this difference masks a deeper commonality: specifically, that mythology and homunculism are both constructed through appeal to the dominant social-organization idioms of the respective times and places in which they were created; Henry Maine long ago proposed that the rise of modernity involved a shift from the domination of social organization through kinship "status" to social organization through corporate "contract".[10] Though varying in particulars, the image of society, and the attendant concern with division of labour, is common to mythology and homunculism; it is the single most exploited image in both realms.

But it would be remiss not to mention the other obvious image common to mythology and homunculism, namely, the image of the body. Like the idea of mind, the body conveys a sense of unity, and, additionally, it bears a recognizable and familiar physical image.[11] It is thus not surprising that mythologies and theologies have frequently inscribed onto the cosmos a body of human-like form, or viewed the human body as a sort of cosmic fractal, a small part that nevertheless mirrors the whole. In homunculism, bodies engaged in interrelated tasks offer an image of unified mental capabilities. Whatever position one ultimately takes on the issue of Cartesian dualism, the body (or coordinated bodies) engaged in executing tasks provides a potent image of mind.

FIRST INVERSION: HOMUNCULIST CONVERGENCE AS THE INVERSE OF MYTHOLOGICAL EXTENSION

As implicit in the previous point, division – in the sense of division of labour – is common to mythology and homunculism: gods functionally divide up the workings of the cosmos; homunculi divide up the processes of brain functioning; the model is that of society or, more specifically, social hierarchy; and the spatial imagery associated with this model tends toward the pyramidal, with a point at the top spreading to a wide base below. The pyramidal image reflects hierarchical subordination (on the vertical axis) and division of function (on the horizontal axis, i.e., the spreading toward the base). As implicit in the discussion above, the spatial image parallels in miniature that of the pantheon of gods that fill and direct the cosmos.

But coexisting with this spatial image, one encounters within homunculist rhetoric images that are *inverse* to the social pyramid – in which, briefly, *extension* in mythology correlates with *narrowing* or *convergence to a point* in homunculism. This possibility is suggested in the fact that the homunculus fallacy is frequently conceptualized as a form of infinite regress, one that is initiated by referring intelligent behaviour to an inner intelligent being (who must also contain such a being, and so on, with the implication that one is moving continuously inward through nestings within nestings). The fallacy is said to inhere in the assumption that such a referral constitutes an explanation. Accordingly the fallacy can be surmounted only by successfully terminating the regress – a goal that homunculism hopes to attain by diagramming mental processing in such a way that the final, stupidest homunculi will be machine-emulatable neural structures.

Now, the capping of infinite series, particularly regresses in time, is a long-established part of cosmological discourse in western philosophy, theology, and mythology. The classic instance is Aristotle's notion of an "uncaused cause," which in some formulations becomes synonymous with the idea of God. A similar concern is often encountered in mythological cosmogonies as well. Polynesian cosmogonies, for example, are among those that account for the origin of the cosmos through the image of procreation; cosmic genealogies are often extremely long, with the earliest elements constructed as if grappling with the threat of an interminable regress. After genealogically tracing the present cosmos far back in time, Maori cosmogonies, for example, sometimes simply terminate the regress by appending the epithet "-the parentless" to the most temporally distant ancestor. Minsky comes closest to systematically interrelating the problems of folk homunculist and cosmological reasoning; he explicitly juxtaposes, for example, cosmological questions such as "*What god made God*?" – a cosmic temporal-causal regress – with the homunculist regress of the "Self-inside-a-Self" (see 1986: 49–50).

Moreover, at least since the time of the Greek atomists, there is a tradition of portraying a bifurcation of the problem of the infinite into two versions that nevertheless retain a connection. That is, the problem of the *infinite* has a counterpart, in some instances an introversive counterpart, in the problem of the *infinitesimal* – the latter giving rise to its own form of capping device, namely the "atom" (literally, "uncuttable" – the smallest possible entity). The notion of a convergence of the outward and inward, of the ex-tending and the in-tending, has sometimes also been accompanied by religious overtones. Augustine's reflections (in the *Confessions*) on the possibility of a human acquiring knowledge of God, for example, suggests in quasi-spatial terms a convergence of the idea of God as that which is most encompassing, with that which is most deeply interior in the human subject – a convergence, that is, of outer cosmological expanse with the most inner recesses of the individual person. Pascal spoke of the relation between the infinite and the infinitesimal: "these extremes touch and join because they have gone so far in opposite directions, meeting in God and God alone" (2008:68–9).[12]

Since spirit is non-physical and needs no space, the only problem in bringing all-powerful spiritual substance to an infinitesimal point is the intrinsic contradictoriness of giving spiritual substance any spatial location at all. In Dante's beatific vision in the *Paradiso* (Canto 28), God is portrayed as a seemingly infinitesimal point emanating light through the cosmos; Dante asks Beatrice to explain, in effect, why the spatial coordinates of matter and spirit are inverse, power correlating inversely with size in the world of spirit rather than in the direct proportion visible in the physical cosmos. It seems not impossible that the homunculist regress of folk psychology might be a source of the Dantean cosmic picture; it would take little more than the projection of the folk homunculist model of mind onto the cosmos to give us such a picture (God as the mind of the universe is among the ancient images rounded up by Copernicus).

In an admittedly loose analogy, homunculism might be thought of as the paired, introversive version of the extroversive anthropomorphic inclinations that give us mythology in its usual definition, i.e., as our outwardly projective cosmic gaze. This would seem to be especially true of the folk homunculism that brings us the homunculist fallacy: the regress of interiority in the form of a self within a self, seemingly without limit and without concern for material or spatial constraint. The brain-science homunculism practised by contemporary AI researchers, however, cannot be so cavalier about spatial considerations. Analytically, the regress of AI homunculism is logical and functional rather than spatial; but the sense of spatial constraint is not absent and seems to add, to the vague interiority of folk homunculism, a new kind of necessity for imagining in miniature and probing the limits of doing so.[13] Mind, according to virtually all contemporary cognitive science, can no longer be thought of as immaterial; hence spatial or at least quasi-spatial thinking is unavoidable. Mind as larger than brain is ruled out; but so is mind as an infinitesimal point, or Dante's perfect inverse of the extended physical world. Mind as brain is a small, interior part; but rather than the mind's full powers mystically converging into an infinitesimal point, we now have an analytical decomposition or division of them that is ultimately capped by the device of machine-emulatable neurons. There may yet remain something of the folk homunculist impetus to think in terms of mind as a whole (in contrast to any of its functional parts) converging to an infinitesimal point; however, in rejecting the image of a "Cartesian theater" – or a single, material place in the brain where "it all comes together" (see 1991:107) – Dennett seems to be warning against such inclinations. Further, in the materialistic universe of contemporary cognitive science, any such inclinations will also be compromised by the materiality of mind (which disallows the infinitesimal) and by the countervailing tendency to conceptualize mind on the social or pyramidal model discussed above, which widens rather than narrows toward the end point.

ை

On this issue of spatializing the mind, two other points should be mentioned. First, it seems that homunculist representation has carried over the spatial orientation that has been attached to mind body/dualism at least since its Judeo-Christian inflection – specifically, that mind (like spirit) is upward tending and matter (or bodies) are downward tending (hence Descartes *descending*). Such spatial coordinates possibly amount to nothing more than a carry-over to organizations of homunculi of the spatial coordinates typically associated with hierarchy (where these coordinates furnish the terms of hierarchy, such as "superior" and "subordinate"), though I suspect they reflect broader inherited traditions in how we conceptualize mind and body.

Second, the quasi-spatial flavour of homunculist reasoning manifests itself not just in the metaphor of little men inside us but as well in the image of making progress by gradually *lessening the gap* – another implicitly spatial image, that of distance – between intelligent behaviour and neurons. One strategy of shrinking the gap is to subdivide functions, although, as we will see shortly, considerable debate exists whether such functional subdivision really brings us nearer to the quality we denote as intelligence. But whether or not this divisional strategy is successful, we have to ask whether the metaphoric spatial ambience itself of homunculist reasoning might not offer a strategy of shrinking the gap, although an illusory one. For, in effect, homunculist reasoning transfers the "God of the gaps" – a classic metaphor for the scientifically as-yet-unexplained – from a cosmic spatial referent to a small, body-internal one. The God of the gaps and the homunculus are both images that arise from mapping a perceived conceptual deficiency onto a spatial grid. In certain respects at least, the homunculist gap is, conceptually, just as "big" (or nearly so) as the idea of a supernatural causality necessitated by a shortfall in human natural science; both gaps are about the creative, mystery, mental stuff of the cosmos. But the gap may *seem* smaller and so much more manageable when portrayed in a local, enclosed, and spatially reduced format.

SECOND INVERSION: HOMUNCULIST OVER-ADEQUACY AS THE INVERSE OF MYTHOLOGICAL INADEQUACY

The second parallel noted above calls attention to the image of social hierarchy in mythology and homunculism; in many ways the social hierarchies of mythology and homunculism reflect one another structurally. But in yet another sense (beyond that discussed as the first inversion), they are reversed: in mythology humans tend to occupy the low end of things, whereas in AI homunculism a human, in the sense of a generic intelligent subject, is at the apex. Even though mythological figures usually have human-like characteristics, most mythologies are nevertheless peopled by beings portrayed as superior to humans. Even the Homeric gods, whose "low" human vices Plato deplored, were more powerful than humans and less subject to human constraints. In most AI talk, the highest homunculus is still within the realm of the human (although the most godlike form of contemporary humanity, i.e., a CEO). The cosmos is bigger than us and so its order and control requires beings bigger and more powerful than us; while in terms of order and control, typically we humans (or at least, to wax totemic, the "top dogs" among us) serve as an adequate image of our own mental capacities.

From this reversal of our place in the hierarchy flows a fascinating inversion between the homunculist fallacy and the theological version of the anthropomorphic fallacy as discussed earlier. In the theological version of the anthropomorphic fallacy, whose original formulation we find in Xenophanes and Plato, the error is said to lie in the fact that the human model is *inadequate*: because the gods are more intelligent, knowing, and powerful than humans, and unburdened by the mundane physical constraints that humans endure, a human model will always prove *inadequate* for conceptualizing the divine. By contrast, in the homunculus fallacy, the error flows from the fact that the human model is *over-adequate*. As a metaphor, the term "homunculus" implies a repositing of the human whole (even though in miniature) for each human functional sub-part. In one sense this is a clever analytical move: the *totum pro parte* logic guarantees in advance that no part will be qualitatively lacking in capacity to "play its part" for the whole. But for the analytical task taken up by homuncular functionalism, that is precisely why the move is analytically problematic: it subverts an analytical decomposition. To have explanatory value, the homunculi team must, at each level, produce higher-level capacities, and at the highest level, intelligent behaviour; but a fatal analytical error has been committed if any of the team members can be shown to already embody the higher-level function that the team is brought in to account for – this would amount to the homunculist fallacy. The problem is not merely the *term* "homunculus" (if it were, a replacement would suffice); rather, the term reflects a problem inherent in any strategy of creating flow-chart nodes whose workings are not understood. In the absence of such understanding, letting the end product, the whole, stand for the part is the only way to guarantee that one has not left out of the part a capacity critical to its function within the whole. Whether the homunculist strategy – specifically that of making progress by decomposing tasks that can be performed by progressively stupider homunculi – will work as a strategy is a matter of debate.[14]

Some of the most interesting issues are tied up directly with the terminology and metaphors. AI homunculists tend to define intelligence in proportion to the range and complexity of tasks particular homunculi "know how" to do; but restricting "intelligence" in this way already removes the concept from a connotation it sometimes has on the level of mind as a whole, where, in some usages at least, it is a quality independent from any particular task. The notion of "stupider" homunculi is similarly problematic. Does stupider necessarily imply more machine-emulatable? Though metaphorical extension is obviously possible, stupid belongs in the first instance to the circle of discourse defined by intelligent and thus carries the notion of intelligence along with it: the stupid affirms the intelligent by connoting an inadequate amount of that quality. In this sense, referring to homunculi as stupid poses the threat of the homunculist fallacy as much as does referring to them as homunculi.

As alluded to above, even the lower homunculi are working not in isolation but as parts of a coherent project. Even menial and repetitive contributions would require "social skills" (delegation, communication, and so on). As noted previously, the tendency in socio-cultural anthropology in recent decades has been to conceptualize social organization as a complex form of symbolic behaviour, standing as much in need of high-level explication as does our capacity for language. It does seem that as AI homunculists decompose functions into homunculi, the "society of mind" (in Minsky's phrase) becomes increasingly elaborate. One danger is that the mind's mystery stuff may be not so much explained as spread around and surreptitiously smuggled in through the image of society – taking advantage of our tendency to unthinkingly presume social-interactional skills for any posited homunculi. This danger is really twofold: on the one hand, such skills would be especially easy to slip in under the radar, since in ordinary life such skills are largely unconscious and tacit; and, on the other hand, it is precisely such overarching, coordinating, teleologically oriented skills that demand intelligence – much more so than the more rote functions often assigned to homunculi. Society is a highly complex, semiosis-laden entity that is easy to take for granted most of the time.

To his credit, Daniel Dennett recognizes some of these vulnerabilities, including the impressionism involved in the judgment of stupider, and the tendency to reapportion intelligence (upward in this case) while nominally moving downward in the hierarchy. At one point he writes,

> One can never be sure ... For instance, the Control component in my model is awfully fancy. It has a superb capacity to address just the right stored information in its long-term memory, a talent for asking M [short-term memory store] just the right questions, and an ability to organize its long- and short-term goals and plans in a very versatile way. This is no homunculus that any AI researcher has the faintest idea how to realize at this time. The ever-present worry is that as we devise components – lesser homunculi – to execute various relatively menial tasks near the periphery, we shall be "making progress" only by driving into

the centre of our system an all-powerful executive homunculus
whose duties require an almost Godlike omniscience. (1978:164)

An "almost Godlike omniscience": why this leap to a level above
the human (given that a functioning human mind sets the upper
boundary of homuncular hierarchies)? Perhaps it is prompted by
a tacit recognition of the analogy between the hierarchical structures of AI homunculi and mythological pantheons. Or perhaps the
attempt to lower, in a sense to dehumanize, inferior homunculi produces a reciprocal or compensatory, reaction: a tendency to super-
naturalize superior homunculi. Also, in the curious semantics of AI
homunculism, and especially for those who regard deities as fictions, a homunculus that is "almost Godlike" may connote a being
that is less analytically threatening than a homunculus that is almost
human-like.

In relation to both of the above problems, the vulnerabilities of the
homunculus image are evident: however diminutive, it is an image of
the whole and thus, as an image, sets up the danger of merely transferring that which is to be explained. Why is an image chosen that
is inherently vulnerable to this charge? It seems that the attitude
toward this image is not entirely consistent. On one hand, there is a
claim that the term itself does not really mean very much. Dennett
says: "Homunculi – demons, agents – are the coin of the realm in
Artificial Intelligence, and computer science more generally. Anyone
whose skeptical back is arched at the first mention of homunculi
simply doesn't understand how neutral the concept can be, and how
widely applicable" (1991:261).[15] That Dennett refers to how neutral the term "can be" is apt, for, in the discourse that has developed
around them, homunculi "can be" used in a continuum of significa-
tions. At one end, they "can" stand for the *present-day reality*, the
starting point of the analysis: we place homunculi at those points
of mental processing that we do not understand, to summarize and
mark our ignorance. At the other end, homunculi "can" represent
the promise or the *end point* of analysis: they are the hoped-for
machine-replicable "units" signifying that our ignorance has been
dispelled. The signification of the homunculus thus spans both the
breadth and duration of artificial intelligence research.

After having at points minimized the significance of the particulars of the homunculus metaphor, Dennett closes *Consciousness
Explained* with an appeal to the importance of specific metaphors;

his concluding words are: "It's just a war of metaphors, you say –
but metaphors are not "just" metaphors; metaphors are the tools of
thought. No one can think about consciousness without them, so it
is important to equip yourself with the best set of tools available.
Look what we have built with our tools. Could you have imagined it
without them?" (1991:455). The dual attitude toward metaphors –
enthusiastic endorsement and elaboration even while gutting or
disowning them – is an issue of importance in evaluating popular
science writing. Metaphors by nature allow the user a certain "wig-
gle room" and a ready-made path of retreat.

As suggested above, in this case at least, the problem is less the
image itself than the shifting demands put on it: the shifting sense
of what, and how much, is implied by the image. Having said this,
however, it should also be noted that the possibility for such shifting is probably enhanced by the choice of a metaphoric image that
is merely a miniature version of its referent. Dennett (1991:455)
regards homunculi as part of a "family of metaphors" that define
his particular theoretical angle. However, the images that we term
metaphors usually involve a major crossing of semantic domains
(life is a journey, love is a firestorm); the homunculus as employed
by Dennett does not do this, and so a question arises whether metaphor is even the best term for this sort of image-play. There is at least
as much homology as analogy; and in terms of the well-known distinction between metaphor and metonym, in many ways the homunculus sits more comfortably with the latter. The homunculus is a sort
of reversed metonym, with whole standing for part rather than part
for whole. The whole that stands for the part has been miniaturized; the representation of partness has been shifted from the more
common register of qualitative incompleteness to that of diminished
overall magnitude. If this is a metaphor, it is not a usual, simple, or
obvious one.

Recall the many other pivotal metaphors we have encountered in
this book so far. Smolin finds the perfect metaphor for his new cosmological theory in the city around him, the city he loves. Copernicus
likens the structure of the universe to a temple, a mind, and a family.
Gould leads us into his views on macro-evolution through the faux-
microcosms of his personal battle with disease and his passion for
baseball. Barrow gives us a scenario of the dawn of humanity based
in part on a literalizing of a pre-eminent mythic symbol, fire. Lakoff
and Johnson start with everyday metaphors and, like Barrow but

more abstractly, they adduce the origins of our humanity, specifically our foundational cognitive mechanisms. Lakoff then follows the family as metaphor into major, contending political philosophies; the political implication of the family is also implicit in Copernicus in pre-modern form: the sun as a ruler on a throne surrounded by his children, the planets. There is a popular notion that what makes a particular image a good metaphor is the degree and precision of the analogy drawn; but this is only part of the story, perhaps the smaller part. At least as important is the kind and degree of human interest and desire that, among the choices available, a particular image will conjure. In varying degrees and ways metaphors are all anthropocentrically motivated; or in other words, metaphors are virtually never "neutral."[16]

FIRST ASYMMETRY: FUNCTIONAL-TELEOLOGIC INTEGRATION OF PARTS INTO A WHOLE IS OF GREATER NECESSITY FOR THE HOMUNCULIST WHOLE (THE MIND) THAN FOR THE MYTHOLOGICAL WHOLE (THE COSMOS)

Despite a good deal of pandemonium in both realms, it seems that when one adds it all up, the societies of AI homunculi are relatively more orderly than the societies of gods portrayed in myths (Plato and his followers regarded with scorn the riotous world of gods and other supernaturals depicted by Hesiod and other myth-tellers). I argue that the difference is a predictable consequence of the inversion of the part/whole relation in mythology vis-à-vis homunculism.

The very concept of mind – the object that homunculism tries to explain – connotes workings or at least end products that are coherent. Mind (or at least particular intelligent behaviours or projects) is the whole, and the homunculi are the parts. The analysis proceeds, in functionally teleologic spirit, to envision a parsimonious organization and division of labour that would accomplish the mind's tasks. The bio-evolutionary assumptions that often accompany homunculist speculations are sometimes visible in the suggestion of an efficiency – far from perfect, but at least workable – honed through evolutionary winnowing.

When we switch to the perspective of cosmology, humans, or human minds, are no longer the whole but the part, while the cosmos is the whole; and when working from the part to the whole in such cosmic context the grounds for assuming functional-teleologic (or any other form of) integration between part and whole seem rather more precarious. To be sure, many if not most traditional cosmologies, and sometimes scientific ones, suppose significant principles of continuity between ourselves and the cosmos around us, or between ourselves and overarching cosmic persons – gods or God. But it is no accident that such portrayals of continuity often have been regarded by skeptics as motivated mainly by anthropomorphic wishes and fantasies, specifically by a desire for a familiar, hospitable habitation. The following passage by a science journalist is one of many that rebuts such desires. "These days, astronomers tell us that perhaps 90% of the universe is made out of matter completely unlike the stuff that makes up our own world. No one knows the exact nature of this 'dark matter,' but whatever it is, it's alien to us. Or more accurately: *We* are the aliens in a universe constructed mostly of other kinds of stuff" (Cole 1999:B2).

It seems that the highest-level concepts – mind, intelligent behaviour, and consciousness – that are relevant to homunculism are, in one respect, not appreciably different from their folk psychological versions. Specifically, such concepts can tolerate only a certain amount of foreign matter before starting to lose their meaning. Examples of foreign matter posited in folk psychology might include "being at cross-purposes," "intrusive thoughts," mental diseases, invasive demons, the idea of multiple "selves," and the lament of the biblical St Paul who complains that he does what he does not want to do while not doing what he wants to do.

Dennett's and Minsky's versions of homunculism both give robust recognition to fractious, competing, even murderous homunculi; but both characterizations contain equally robust depictions of unifying mechanisms that bring much of the pandemonium into final coherence. While Dennett's "pandemonium of homunculi" model lacks the efficient organization that one might have expected to triumph in the Darwinian battle, the achievement of at least a degree of overall workability is affirmed by the very fact of survival within evolutionary process. Contrast this to the inverse situation, in which humans or human minds are the parts and the cosmos is the whole. The logic of parts and wholes in this case seems to dictate virtually no (or just short of no) constraint on the whole to be like humans or to have anything to do with us – although many people find the idea of a universe alien to rather than homologous with us, or the idea of a God that is "wholly other," to be intolerable. Indeed, in the idea of the alien universe we arrive at a minimal and ambiguous meaning of the term "part," i.e., to lie non-functionally within.[17]

It must be acknowledged that the logic inherent in part/whole relations is not the only factor operating to determine the final shape of a homunculist or cosmological theory. Especially in cosmology we encounter a variety of culture-specific doctrines that influence the extent to which the human microcosm and the cosmic macrocosm are held to be homologous. As noted previously, Maori and other Polynesians portray the cosmos in the form of a genealogy, in which humans and other living things are descendants of the first cosmic parents, Sky and Earth, who together with the unfolding genealogy of their descendants, form the borders of the cosmos. Humans and other living things will thus possess a degree of continuity of substance with the cosmos as a whole by virtue of shared ancestry.

But not all cosmologies end up affirming such continuity. Indeed, the heroic vision of ourselves as the odd bit of foreign matter in an indifferent cosmos is a favourite scenario of modern existential angst, encountered in a variety of literary and popular scientific contexts, some of which will be considered in the following chapter. Among the motivations for modern-day calls for returning to myth is a desire to restore continuity and harmony of self and cosmos over against the rupture allegedly introduced by the dominance of science in our era.

But while restoring the continuity of humanity and the cosmos for some amounts to an anti-science stance, this is not so for all. I refer the reader to a work of cognitive science, with a foreword by Daniel Dennett, that has recently received a good deal of attention, Zoltan Torey's *The Crucible of Consciousness* (2009). Announced early in this work and reaffirmed throughout is the claim that cognitive theory must define itself over against the mindset of myth or mythology, which Torey links with animism, fantasy, illusion, parochialism, dualism, comfort-seeking, distortion, self-deception, mystification, and anthropocentric narcissism. Not the first to do so, Torey finds in the evolution of the mind's powers of abstraction the source also of its inclination to self-delusion: "the abstracting process may become a mythogenic cauldron for making up pseudo-explanatory schemata that ignore or distort reality" (2009:75). However, in his final chapter Torey addresses the question of the cosmic significance of the human species; and here he introduces a mystic scenario from Cabala, telling of emanations of divine light that spilled over when God withdrew into Himself, leaving "sacred sparks," which, through virtue, humanity strives to recover. Torey says: "The aspect of particular interest in this scenario is that it assigns a definitive role to humanity and that the role represents a measure of symmetry between humans and the universe, in other words, that there *is* something that humans can do for the universe and that this potential contribution is significant. The idea is imaginative and dignified and at variance with the usual run-of-the-mill mythologies" (2009:199).

In the language of the singularity, the Big Bang, black holes, negative entropy, biospheric viability, and collapsing the quantum wave, and alluding to our intermediate position between large and small scale (the chapter title is "Between the Quantum and the Cosmos"), Torey sets out an elaborate if obscure vision of the human mind as having an "instrumental" role in the evolving cosmos; the mind is "the 'means' for the promotion of what the universe is all about: the singularity's struggle to achieve self-reconstitution and oneness" (1009:201). He insists that, unlike the strong version of the anthropic principle, his theory is not mythic and not condemned to illusion. As with Stephen Hawking's "mind of God" comment, one wonders how Torey's coda, at odds with the tenor of most of what has preceded it, should be taken. At face value it offers a scientifically inflected scenario of human functional-teleologic integration into the cosmos that rivals cosmic visions encountered in many traditional myths, and which is worthy of comparison with the functional-teleological integration that binds a society of homunculi into a body's mind.

ᘉᘏ

Even if one were to accept Torey's claim that his theory itself is not mythic, there is much that is undeniably mythic in the way it is announced. Given the anti-mythic rhetoric he maintains in the lead-up to the final chapter, it is understandable that Torey must rhetorically differentiate his favoured myth – the Cabalistic one that adumbrates his theory – from "run-of-the-mill mythologies." Notably, however, Torey does not identify any of these "run-of-the-mill" mythologies; and from the standpoint of comparative mythology, I must protest that his comment is absurdly parochial (or mythic, in his terms). All of the themes he finds – the breakup of an original whole, the scattering of remnants, the search to recover these, and the onus on humans to bring cosmic processes to fulfillment after the creators withdraw – are recurrent themes in world mythology.[18] Torey legitimizes his favoured myth, while putting down others, through the implication that his myth anticipates the findings of science. Once again, he is not the first to employ such a strategy; indeed, this method of legitimizing a myth amounts to a new variant of the allegorical tradition that began with Greek philosophers, torn between wanting to dismiss myths and wanting to save them (or at least some of them).[19] Reading mythology as prophecy is as old as mythology itself.

In recognition of the possibility of such cosmological doctrines, my argument about constraints determined by part/whole relations can best be summarized as follows: despite room for considerable internal discord or pandemonium, the part/whole relation characteristic of homunculism ultimately also requires a degree of functional-teleologic integration of parts into a whole (that is, of mind parts – homunculi – into mind); whereas in cosmology such functional-teleologic integration (i.e., of humans as parts into the cosmic whole) is not required, though it is sometimes affirmed – often in mythology, but also occasionally in the name of science. Both homunculism and mythology offer devices, minimally forms of imagery, that keep us from having to stray far from ourselves. For homunculism, the whole is a real-world-sized human, while the parts are little humans; in the cosmos of many mythologies, and also in the cosmos of science according to a thinker like Torey, real-world humans form constituent parts while the cosmic whole is an oversized human – in the form of an anthropomorphic creator-deity and/or in the form of an oversized humanity, owing to the work of real-world-sized humans in bringing it to completion.

SECOND ASYMMETRY: THE UNIDIRECTIONALITY OF THE ARROW OF TIME AND ITS CONSEQUENCES FOR AI HOMUNCULISM

Immediately after discussing a contemporary cognitive theorist's temporal reversion to an older mythological idea may seem an odd moment at which to assert the unidirectionality of the arrow of time. The history of human thought is full of cyclic events of rediscovery, revival, and re-awakening, whether by strategy or coincidence.

Cyclicality in intellectual history is of special interest to a mythologist, just because the very term "myth," through its association with another term, "ritual," tends to connote the cyclic. But having said this, it is also necessary to acknowledge a certain fixity in the history of thought: we can say, for example, that viewed within the frame of the everyday world in which thinkers develop their ideas, Marvin Minsky has the opportunity to build on Hesiod in a way that Hesiod did not have to build on Minsky.

My focus in this section is a powerful schema of the development of science that has been with us for some time and that was given a particularly forceful formulation by the early nineteenth-century social thinker Auguste Comte. This scheme figures importantly in the nineteenth-century criticisms of mythology, specifically in the scientific version of the anthropomorphic fallacy. The same scheme appears often in portrayals of the long-term project of AI homunculism. The asymmetry upon which I focus stems from the *order of appearance* – specifically, the different locations in the timeline of science – occupied by the anthropomorphic fallacy and the humunculist fallacy.

Comte's vision has exerted, and continues to exert, a strong background influence on contemporary consciousness. He articulated a cosmic scheme of the growth of science through a process in which academic disciplines were gradually brought under its purview. The main points of Comte's scheme are as follows:[20]

- Academic disciplines develop in an order that reflects a hierarchy of complexity of different types of phenomena.
- Complexity is determined by the number and type of natural laws operating in different kinds of phenomena. Each more complex type of phenomenon presupposes the laws of less complex phenomena, and superadds a new level. The hierarchy of disciplines ranges from astronomy, with the fewest and most basic laws, through biology, to social and moral sciences – each adding a new level. These respective disciplines thus unfold logically and sequentially, according to complexity envisioned as levels of superadded laws.
- The development of any discipline amounts to that discipline becoming scientific. The transition to science is marked by passage from a perspective dominated by spiritual/personal agencies to one dominated by recognition of impersonal laws of nature.

- On the whole, science is progressing because, in regular and orderly fashion, one after another class of phenomena is being brought under its purview.

೧೦

The Comtean scheme portrays humans both centrically and decentrically. Although seeking to replace mythico-religious conceptions, the Comtean hierarchy offers its own version of humanity as the crown of creation, figured through complexity of laws.[21] Moreover, stemming from this placement, there is a sort of anthropocentric directionality in science, which followers of Comte love to dramatize. Science starts with the distant planets and marches – inexorably – toward us, the final holdout: it wants even our interior lives. In the Comtean scheme we are in a sense special: the march is oriented toward us, and we are its highest prize. Yet there is also a devastating decentring implication in the assumption that each realm will relinquish the conceit of being possessed of a substance that puts it beyond the march. While being confirmed as the summit of the hierarchy of nature, we are reduced from standing above, to being a part of.

The directional schemata associated with the triumphant view of science are fascinating in themselves. The first and most obvious is the theme of upward directionality portrayed in most eighteenth- and nineteenth-century depictions of the growth of science and the progress of culture. But as the comments just above suggest, this was frequently accompanied by a sense of science progressing *inwardly*, from distant planets toward us humans: the march is upward *and* inward. We face the choice of immediate surrender or strategic retreat – a strategic contraction of the borders of the redoubt.[22] Finally, homunculism suggests an interesting twist to this inward directionality: a sort of Trojan horse move, or hollowing out from within.

Elements of Comte's scheme figure powerfully in nineteenth-century critical treatments of mythology. Remnants of anthropomorphism and animism were assumed to reflect scientifically unevolved academic disciplines, in which personalized forces had not yet given way to the idea of natural law. The unfolding of the logically more elemental and thus chronologically prior sciences, especially astronomy, provided a luminous example of anthropomorphic understandings of the universe being replaced by impersonal, rigorous, predictive science.[23] As suggested above in Tylor's comment, the exemplary case was the development and success of Newtonian planetary physics. Although in important ways consonant with Tylor, Minsky (2006:13ff.) interestingly blames an aspect of this model – that psychology should imitate physics – for retarding the study of the mind over the last three centuries. Minsky argues that instead of attempting, in Newtonian fashion, to reduce mental processes to a few simple laws, the study of the mind must make its topic more complex by breaking capacities into smaller parts – the strategy, in other words, that we have been considering in this chapter.

⌒⌒

The Comtean schema shows up in a variety of ways and degrees in homunculist debates. Importantly, according to this schema, the study of the outer cosmos becomes scientific before the study of the human mind undergoes the same transition; this order-of-occurrence gives rise to several asymmetries in the way we conceptualize the science of the mind vis-à-vis the science of the planets. I will here discuss two dimensions of asymmetry, one very practical, the other more abstruse.

The more practical dimension of asymmetry is that one can defer taking a final stance on ontological dualism when talking about the planets, but can no longer do so when talking about mind. That is, one can concede a monistic, materialistic science of the planets while holding out for dualism in the realm of mind, but not the reverse (or at least much less easily the reverse, since the mind is the favourite site for locating the distinctive "*cogitans*" substance, the last bastion vis-à-vis the advance of science). One can imagine two epistemological soldiers fighting as comrades in the battle for understanding the planets through materialistic science, only to discover that they belong to opposing sides when they return home to fight battles over the nature of mind. They return home to discover that one of them was fighting against anthropomorphizing planets because planets do not contain the mind stuff that humans do; the other was fighting because the existence of distinctive mind stuff anywhere is an illusion – the planets were merely the first campaign.

The second, more abstruse and intriguing, asymmetry occasioned by the order of unfolding in the Comtean scheme, stems from the fact that later science has prior science to draw upon for inspiration and lessons on how to become scientific. The heroic saga of the growth of science and the idea of myth as a sort of protoscience or step toward science are so broadly disseminated that it is not at all stretching matters to suggest that among the incentives to the creation of AI homunculi might be an impetus toward imitative creation of mythology – not toward mythology as an end in itself, but rather toward mythology as a necessary step to an end, namely, the development of science. It would be a little like the dilemma of orthodox Marxists asking whether they should promote capitalism in order to hasten its inevitable demise. Little pantheons of homunculi, viewed in the long term, offer a roundabout way of saying that the mind is like the physical cosmos, which was also once thought of in terms of pantheons. All that remains is the replacement of these personalized understandings (they can be picked off, one by one) with depersonalized ones. Mediated by the Comtean scheme, mythologizing is the first step toward demythologizing.

The presence of such an impetus is suggested in several ways, most interestingly in the recurrent theme that the homunculi, while necessary now, have no long-term career prospects. Dennett says that "all homunculi are ultimately discharged" (1978:124). Baddeley hopes "in due course to make the homunculus redundant, and pension him off" (1995:760). Fodor characterizes the little man as a "representative *pro tem*" (1999:47).[24] Seemingly, homunculism and Cartesian dualism will end at the same moment, when we arrive at Neuronopolis and, like mythological gods who have handed over their creations to lesser powers or been driven out by them in a sort of inner-mind *Götterdämmerung*, we find no place there for ourselves. A century ago classical comparative mythologist James Frazer spoke of the "deserted" cosmos that was science's legacy *except for* our own consciousness. In his recent work, *Sweet Dreams*, Dennett stands ready to deliver the final eviction notice: "[a] good theory of consciousness *should* make a conscious mind look like an abandoned factory ... full of humming machinery and nobody home to supervise it, or enjoy it, or witness it" (2005:70).

EPILOGUE
ALL OF ME: THE HOMUNCULIST FALLACY AND THE
TOPOGRAPHY OF ACADEMIA

The resonances between the issues implicated in the homunculist debates and those familiar in the study of mythology are palpable enough that in the course of considering them I have not been able to avoid asking questions about our present-day topography of know-ledge. What factors eventuate in these topoi being academically so distantly situated vis-à-vis one another? It was by a rather tangential route (namely, science education films by Frank Capra) that I came to connect them. And, to be a bit less subjective, I note that Guthrie's detailed survey of anthropomorphizing has little that is specifically relevant to the homunculist debates amidst the large number of ref-erences directly relevant to mythology. The closest Guthrie comes to homunculism is a brief mention of scientific language in which cellular-level structures are anthropomorphized (more on this at a later point).

The topography of intellect – the ways in which we arrange and classify our world of knowledge and theory – perennially presses on us the need for critical examination. I find the topographical issues relevant to mythology and homunculism no less absorbing than either focus considered directly; and therefore would like to summar-ize, once again in a provisional, schematic form, the principles most relevant to the classification of homunculism vis-à-vis mythology.

A. As noted above, there is a background temporal/spatial distribu-tion at work, evident in the popular tendency to think of mythol-ogy as peopled by gods and ancient heroes, while homunculus talk tends to be peopled by moderns, such as corporation functionaries (CEOs and the like). The surface feel thus might seem quite different, among other things pulling toward very different epochs of time. But, as noted above, this distinction is superficial, and underlain by the more important commonality of society as the core image (the second parallel above).

B. Perhaps reflective of the background distribution just noted, there are the placements within inherited disciplinary taxonomies. Myth-ology belongs to classics, literature, and other humanistic fields, while the homunculus debates and related issues belong to psychol-ogy, cognitive science, and philosophy of mind. We cannot do away with disciplinary classifications, but these are also potential sources of misleading disjunctures.

Those interested in the study of myth and folklore encounter, in these two terms, yet another problematic distinction. Though myth is taught in some academic folklore programs, there is also a sense of tension between folklore and myth, one stemming in part from the

186 The Ancient Mythology of Modern Science

way the two terms align with the temporal/spatial distribution noted earlier. Myth designates the spatially very distant and the intellectually very wrong, while folk designates the temporally/spatially closer (the things one's grandparents believe) and contemporary everyday rough-and-ready opinion (vis-à-vis specialized knowledge). It is thus no accident that cognitive science invokes folk (as in "folk psychology," "folk taxonomy," etc.) to mean the everyday non-specialized, and myth to designate falsehood. Folk understanding is seen as worthy of engaging in debate, while myth is assuredly wrong.

C. Another factor that has tended to work against introversive and extroversive projection being explored within a common framework is the theoretical tradition that has developed around the idea of psychological projection. Specifically, projective theories have tended to develop in terms of unidirectional, two-grid models. One grid is the source grid; the other is the recipient grid, the screen onto which images are projected. The models tend to be constructed such that one examines only one directionality of projection at a time. The projective grid for Freud was the set of dyadic relationships that clustered in the nuclear family; the projected grid was the cosmos.[25] For Durkheim and his students the projective grid was society, and the projected grid was also, as for Freud, the cosmos. It is worth exploring alternatives to this tradition; for example, Robert Levy, in his *Mesocosm* (1990), explores ways in which humans portray cities as suspended between both larger and smaller structures that must be reconciled, thus offering an alternative to the two-grid model. And it must be added that although the Freudian analytic model at any moment tends to be unidirectional, there is an intriguing sense in which projection is also bidirectional. Specifically, the dyadic relationship (say, the relation of parent and child) that is postulated to be the source of cosmic projection in myth and religion (where the parent/child relation becomes the god/human relation), simultaneously is projected, by the psychoanalytic theorist, introversively in the form of components of the psyche (specifically as the relationship of superego to ego and id). Humanity's transcendent visions are thus recast immanently, in the form of a theory of the psyche. Instances of such simultaneous bidirectionality deserve further attention.

D. Once more perhaps reflecting the background distributions, there is a powerful tendency, noted at the outset, to dichotomize between imagery invoked in myth, on one hand, and imagery invoked in literature and science, on the other. The thinking behind this seems to be that the former involves literal identities, or in other words "beliefs," which, considered from a scientific standpoint, are errors (under the supposition that ancient and allegedly primitive peoples erroneously believed that planets and stars *really were* people). Contemporary artists and scientists are said to self-consciously invoke figurative language not in error but, by contrast, in metaphors with heuristic and artistic value. Behind this dichotomy, myth vs. metaphor, lies another dichotomy, and the most powerful of all: falsity vs. truth. These two dichotomies historically have come to be aligned, and at this point we must ask whether the alignment might be too convenient.

As noted earlier, nineteenth-century mythologist E.B. Tylor was aware of, and troubled by, the fact that the romantic poets of his time invoked the same sorts of images that were the coin in trade of ancient solar mythologies. The commonality in imagery – especially in personifications of nature – in texts from ancient and modern peoples threatened to overturn Tylor's notion of a unilinear evolution of culture from the starting point of myth (i.e., primitive error) to a final state of science and high art. Tylor sought to blunt the threat with a formulation to the effect that what savage peoples do in error civilized people do as art.[26] The net effect was to send myth and metaphor down different analytical tracks, defined by different attitudes of mind and epochs of time and in terms of diametrically opposed relations to science, art, and truth. But this is only one of many possible ways of classifying these topics. An alternative classification, one which de-emphasized the precise nature of the mental engagement (i.e., whether literal or figural) and foregrounded instead the kinds of imagery chosen and how it is structured, might be quite revealing.

Homunculism, of course, falls within the range of metaphor, and thus out of the range of myth, according to the long-standing dichotomy promulgated by Tylor. But this dichotomy itself can also be challenged. First of all, the dichotomy of myth as literally believed and metaphor as literally disbelieved may not always be as neat as Tylor would have it. The fact is that in many of the studies of mythology in which it would have been possible to investigate such issues, they were not broached in any depth – no doubt because the regnant theory proclaimed the answer in advance. It is entirely possible that traditional mythologies were subscribed to by their original adherents with multiple and complex (e.g., tentative, ambivalent, shifting, context-bound) attitudes.[27] Similarly, while it is clear that homunculists are not looking for actual little men, I have noted above some complexities that suggest that this image functions in ways that strain the notion of metaphor.

For yet another reason, much of the scholarship on metaphor in the present era tends to support the separation between myth and metaphor. Many specialists in metaphor (including Lakoff and Johnson, considered in chapter 4) want to show the centrality of metaphor to mental life in general, something that can be accomplished by showing how permeated everyday, ordinary thinking is with metaphor. To bring myth into the argument – laden with the connotation of ancient and exotic thought, now superseded – is not particularly useful, and could even be counterproductive, for this goal. Thus there is one more reason to keep work on myth and metaphor separate, even though the same images tapped for the "metaphors we live by" frequently have also been mined by myth.

E. Importantly, our ordinary, everyday way of experiencing our bodies, our "selves," our "minds" – our folk psychology on these matters – itself may be a source of resistance to bringing the introversive projection of homunculism into the same category as the extroversive projections normally termed anthropomorphic. Homunculism is said to be rooted in folk psychology: we have a sense of an interior self or voice that is more deeply self than are our extremities. And yet the phenomenology of the body also encompasses a countervailing tendency to insist that any part is irreducibly the whole. A sort of hypostatized metonymy is reflected in my feeling that "it's *me* you're cutting!" when the doctor makes a small incision on an extremity; the feeling extends even beyond the body to encompass what early anthropologists thought of as an allegedly primitive doctrine of the individual's "contagion" of his/her "appurtenances" – and which we now refer to as an individual's "personal effects." The body has its own reasons; to imply "all of me" in a part is felt as a principle of integral identity, not as a foisting of human identity on something non-human (the latter is what anthropomorphizing amounts to).[28] Only when we reach the *remotest* extremities of our physical structure does the resistance let up and allow us to think of this projection as anthropomorphism. As if in confirmation of this principle, I am intrigued to note that in his survey of anthropomorphism, Guthrie makes no mention of the AI homunculism debates I have been discussing; yet he does include, as examples of anthropomorphism, biologists referring to the way that the cells of our body talk to one another (and other such anthropomorphic descriptions of processes at the microscopic level of DNA and genes) (1993:171–2). Perhaps it is the everyday familiarity of homuncularizing that precludes thinking of this process as anthropomorphizing: the "little voice inside" implies one's deepest self, not a projection of one's self onto something other. But when we reach the level of cells, DNA, or genes, we have reached extremities of our structural composition unfamiliar in everyday experience and on which we are foreigners to ourselves; at this level the sense of anthropomorphizing – as when we say that our cells talk to one another – predictably reappears.

An analogous dilemma – are we anthropomorphizing or extending our self-concept legitimately? – is confronted in the realm of primate research.[29] We think of non-human primates as especially similar to ourselves, and science affirms the overlap and shared ancestry. Thus, projection of our thoughts, feelings, and motives onto chimpanzees and gorillas in analyzing their actions may seem less like anthropomorphizing than does the same projection in the case of a cat or a clock. If the dilemma in the case of homunculism stems from projecting ourselves onto beings that are *parts of* ourselves, in the case of primate research it stems from projecting ourselves onto beings that are *partly* ourselves.

F. As a final point, I note that there would be something particularly absurd in referring to homunculism as the anthropomorphizing of mind. For, as earlier noted, it is precisely mind that we project *onto* non-human objects when we anthropomorphize them. The idea of anthropomorphizing mind thus has a kind of intrinsic tautology. Now since tautology or redundancy is the very error charged by the homunculist fallacy, apparently we face a choice between two ways of representing the threat of tautology in accounting for mind. Specifically, we face a choice between a directly and obviously tautological formulation (the anthropomorphizing of mind), on one hand, and, on the other, the threat of tautology portrayed, rather, as the threat of infinite regress (i.e., homunculism). Such dilemmas are not limited to AI homunculism, for the spectre of introversive anthropomorphizing arises wherever mind or motive are referred to interior "parts," an old venture that, in new varieties, now captures our imagination.

Figure 6.1   Earthrise. Earth as photographed from the moon by Apollo 8.

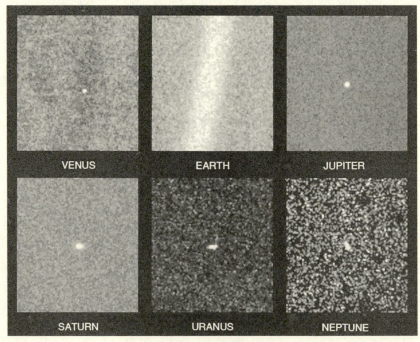

Figure 6.2   Family portrait. Earth and other planets as photographd by the Voyager probe as it left the solar system.

6

Once More, with Feeling! The View from the Moon and the Return of the Copernican Revolution

Science offers little in the way of cheap thrills.

Carl Sagan, *Pale Blue Dot* (1994:365)

Long a topic of interest for history and philosophy of science, the Copernican Revolution more recently has become a favourite symbol for popular science writers. This symbol is used in two main ways, both of which shade easily into the mythic. First, the Copernican Revolution is invoked to mark a point of grand transformation between anthropocentric and objective thinking, between mythic darkness and scientific light. Second, many portrayals of the Copernican Revolution heroize this protagonist and ritualize his deed. The Copernican Revolution designates an alleged shift in worldview that took place in the past, but also a sort of personal microcosm of that event – a rite of passage that any mind aspiring to science must undergo. The term "myth" in some usages becomes a near-synonym for pre-Copernican thought – a definition that is maddeningly inadequate even though refreshingly concise.

Thomas Kuhn's widely read mid-century work, *The Structure of Scientific Revolution*, portrayed the Copernican Revolution as the exemplary case of "paradigm shift," an idea that was taken up across academia and even in non-academic discourse: it is as if there are two kinds of revolutions – the ordinary kind and the paradigmatic, prototypic, Copernican kind. In the post-Kuhn era, both Kuhn's formulations and the idea of a Copernican Revolution more generally have been subjected to a number of challenges within the history of science (concerning, for example, the question of just how much of the credit should go to Copernicus).[1] But in science writing the challenges pale in comparison to the florescence of the Copernican Revolution as an instructional and inspirational image.

This chapter focuses on a specific recent revival of Copernican musings, one prompted by images of Earth photographed from the moon in the course of the Apollo moon missions of the 1960s and '70s. These musings have a distinctly late twentieth-century signature, apparent not only in the technology of space travel but even more importantly in the accelerated technologies of image processing, allowing real time visual images from the moon to be broadcast on Earth. Before turning to this specific focus, however, I would like to set the stage by offering a brief, selective tour of some of the other incarnations of the idea of the Copernican Revolution.

I will begin with a deceptively elementary version of the idea, that offered in Steven Weinberg's *The First Three Minutes*, a work that provides a synopsis of theories of Big Bang cosmology as they stood in 1977. The book's opening sections contain two terse references to the Copernican Revolution. One of these rehearses the most elementary and immediate humanistic implication commonly drawn from it: "ever since Copernicus we have learned to beware of supposing that there is anything special about mankind's location in the universe" (1984:20). The other is more intriguing: "We would expect intuitively that at any given time the universe ought to look the same to observers in all typical galaxies, and in whatever directions they look ... This hypothesis is so natural (at least since Copernicus) that it has been called *the* Cosmological Principle by the English astrophysicist Edward Arthur Milne" (1984:18). This passage suggests the *depth* of transformation connoted by the alleged revolution; it implies that what is *intuitive and natural* differs between the pre-Copernican and Copernican worlds.

Weinberg's introduction opens with a synopsis of a Norse myth (from the *Younger Edda)* that accounts for the origin of the cosmos through a story about a giant and a cow, an episode that gives us "not a very satisfying picture" (1984:1). Weinberg invokes a comparison between the Eddic myth and scientific cosmological theories a number of times through his book, thus exemplifying the trend to which I alluded above – that of locating the opposite of the Copernican world in the world of myth. "We are not able merely to smile at the *Edda*, and forswear all cosmogonic speculation – the urge to trace the history of the universe back to its beginnings is irresistible. From the start of modern science in the sixteenth and seventeenth centuries, physicists and astronomers have returned again and again to the problem of the origin of the universe" (1984:1).[2]

Although sensitive to "not offend religious sensibilities" (1984:1), Weinberg nonetheless refers throughout to the problem of the "Genesis" of the cosmos (repeatedly capitalizing the term). Although Weinberg says little else of direct relevance to the Copernican Revolution, he does sketch, in the closing passage of *The First Three Minutes*, a terse, humanistically compelling vision as if to make up for our decentred position in the cosmos. I will return to Weinberg in order to consider this and other such compensatory visions, after considering some of the more elaborate depictions of the Copernican Revolution offered by other science writers.

Two types of elaborations in particular stand out. One involves a litanizing of major revolutions in thought, with a tendency toward groupings of three; the other involves an attempt to derive a more general "thinking lesson" from Copernicus.

COPERNICAN TRINITIES

In its more elaborate incarnations, the Copernican Revolution is often presented as a revolution that precipitated other like-minded revolutions; indeed later scientific revolutions are sometimes assimilated to this one. Let us reconsider, for example, the epigraph from astronomer John Wheeler that opened chapter 4 of the present work:

> That quick dismissal of the idea of a universe without life was not so easy after Copernicus. He dethroned man from a central place in the scheme of things. His model of the motions of the planets and the Earth taught us to look at the world as machine. Out of that beginning has grown a science which at first sight seems to have no special platform for man, mind, or meaning. Man? Pure biochemistry! Mind? Memory modelable by electronic circuitry! Meaning? Why ask after that puzzling and intangible commodity? "Sire," some today might rephrase Laplace's famous reply to Napoleon, "I have no need of that concept." (1988:vii)

This passage assimilates to Copernicus scientific developments of later times (and that he could not have known about, such as electronic circuitry). The passage is interesting for another reason also.

Specifically, the most robust versions of the Copernican Revolution are typically put forward by writers who think our cosmic insignificance is in need of emphasis. By contrast, Wheeler's statement introduces a work that promotes the "anthropic principle," which challenges the idea that it is necessarily unscientific to speak of humans as holding a special place in the cosmos.[3] Wheeler's statement thus illustrates the fact that robust elaborations of particular ideas are also useful to those who wish to challenge them; such elaborations can serve as "straw men" for polemical treatises.

Some of the patterns evinced in Wheeler's comments are broad spread in contemporary writing: particularly so is the inclination to overlay Copernicus with later revolutions, with the implication that all of these revolutions are like-minded, and sometimes also with the implication that each subsequent revolution strikes a new blow against the human ego. A potent source of this particular strategy is Freud's proclamation of three great blows that science has struck against human self-infatuation, respectively by Copernicus, Darwin, and the insights of psychoanalysis (Freud 1964:284–5; the full passage is quoted in chapter 1). As one small indicator of the power perceived in this trinitarian formula, I note that of the authors at the centre of the five case studies in this book (chapters 2 through 6), four – Barrow, Gould, Dennett, and Sagan – at some point give us a variation on the formula (these will be considered in the course of the discussion below). It is interesting that this formula, so beloved of science popularizers, has as its source of propagation not a scientist (in any sense that would be accepted today), but a psychoanalyst, one who was intensely interested in religion, and who was himself a trinitarian of sorts (id, ego, superego). The parallel of Freud's formula to the Christian Trinity should not go unnoticed: the first person, the Father, is associated with Creation and the design of the universe (the concern of Copernicus); the second person, the Son, is associated with the place and fate of our species in the cosmos (Darwin), and the third person, the Holy Spirit, biblically a being of shifting iconicity, is divine energy in the here and now (the passionate science writer's vision of his/her own field).

A number of writers have offered modifications to Freud's litany. Thomas Kuhn (1985:4) adds Einstein to Copernicus, Darwin, and Freud. As I noted (in chapter 3), Stephen Jay Gould picks up on Freud's triadic litany, re-dubbing the "blows" as "Freudian bullets," with the implication that once we stop denying and bite the bullet we will be better off. Gould invokes the Freudian trinity in order to claim that a step has been left out: to the discoveries of Copernicus, Darwin, and Freud should be added the insight – which he credits to his own academic field of paleontology – that cosmic time is not, as the Biblical view has it, co-extensive with humanity. Gould pronounces this discovery "the temporal counterpart to Copernicus' spatial discoveries" (1997:18). As does Weinberg, Gould repeatedly opposes the post-Copernican world to the world of mythology (see chapter 3).[4]

Despite its merit, it is unlikely that Gould's revised litany will supplant Freud's original, which is far more graceful. One reason that Freud's original formulation is so powerful is that it incorporates three other broadly recognized patterns; and Gould is out of step with all of these. First, Freud's formulation recapitulates a broadly acknowledged threefold hierarchy in the organization of matter in the cosmos, the earlier-mentioned three great leaps in nature: from nothing to something, from something to life, from life to mind (or, more tersely, inanimate matter, life, and mind). Second, as alluded to in the previous chapter, nineteenth-century social evolutionists (who directly influenced Freud) portrayed the progress of science spatially as a movement both *upward* and *inward*; last of all science will take root in the human psycho-social and moral spheres.[5] Freud's triad perfectly captures this sense of evolution's *inward* march, with Freud's own science of mind as its culminating point; Gould's fourth bullet is out of step. Third, in advancing a four-bullet scenario, Gould violates what folklorist Axel Olrik called the "law of threes." In Western literary, mythico-religious, and folkloric tradition, the sign of completion of a quest, the number of obstacles to be overcome, is three.[6]

ᕮᕭ

Variations continue to arise around Freud's trinity; one such did as I started to collect material for this chapter – in computer guru Stephen Wolfram's much-anticipated and controversial book *A New Kind of Science*. Although the tome is massive, the section devoted to "Historical Perspectives" amounts to little more than a new variation on Freud's schema:

It would be most satisfying if science were to prove that we as humans are in some fundamental way special, and above everything else in the universe. But if one looks at the history of science many of its greatest advances have come precisely from identifying ways in which we are not special ...

Four centuries ago we learned for example that our planet does not lie at a special position in the universe. A century and a half ago we learned that there was nothing very special about the origin of our species. And over the past century we have learned that there is nothing special about our various physical, chemical and other constituents.

Yet in Western thought there is still a strong belief that there must be something fundamentally special about us. And nowadays the most common assumption is that it must have to do with the level of intelligence or complexity that we exhibit. But building on what I have discovered in this book, the Principle of Computational Equivalence now makes the fairly dramatic statement that even in these ways there is nothing fundamentally special about us. (2002: 844)[7]

Daniel Dennett's recent work, *Darwin's Dangerous Idea*, offers another variation on Freud's triadic litany, one which like the Wolfram version, gives the third slot to research involving the modern electronic computer. Dennett's work also alludes to the previously mentioned theme of the inward advance of science. "No wonder, then, that there should be so much antagonism to both Darwinian thinking and Artificial Intelligence. Together they strike a fundamental blow at the last refuge to which people have retreated in the face of the Copernican Revolution: the mind as an inner sanctum that science cannot reach" (1996:206). The idea of a series of strategic retreats – of a contracting perimeter around the final bastion of human cosmic specialness – is the other side, the anthropocentrically "defensive" side, of the idea of the inward advance of science.

In Dennett's and Wolfram's formulations, the only missing member of Freud's trio is Freud; he has been replaced by computer emulation of mental processes, a field that, as noted in the previous chapter, offers an updated emblem of the promise of a science of mind. The replacement of Freud is made easier by the fact that Freud did not actually name a specific hero to the third slot – the mind slot – but, rather, pointed to the contributions of a field of inquiry, though one in which he himself played the central role. Judging by Gould, Wolfram, and Dennett, this practice has become part of the formula, for these three, too, modestly point to a field of inquiry rather than a specific individual – although it happens that each names his own specialty. John Barrow, whose cosmology we considered in chapter 2, in the 2005 expanded edition of *The Artful Universe*, also adds a new variant to the trio, naming his own expertise and interest, the discovery of universal constants of Nature – a "super-Copernican step" – as the third great advance, following those of Copernicus and Darwin (2005:220).

Dennett's portrayal also aligns the Copernican Revolution metaphorically with another potent image of transformation, specifically, the idea of the loss of an original paradisiacal state (with overtones of the story of the Garden of Eden) – the passage from blissful ignorance to uncomfortable but truthful understanding. He opens his first chapter with the Sunday-school song that asks, "Tell me why the stars do shine" and answers, "Because God made the stars to shine." He interrupts the idyll with, "and then along comes Darwin and spoils the picnic" (1996:17). But Dennett then quickly challenges this terse interpretation of Darwin, asking, "Or does he?" He takes up the task of persuading us of the nobility of the Darwinian vision, thus offering us an adult's compensation for the secure but childish vision we must surrender.

Gould's and Dennett's versions of the Copernican Revolution are robust in form and categorical in tone. Both of these writers repeatedly juxtapose their perspectives with those of myths or mythico-religious traditions. Yet at the same time, Gould's and Dennett's visions of the progress of science are themselves highly mythologized. Both are constructed around the type of characters that mythologists call "culture heroes," Promethean figures that are said to have introduced to humans pivotal, transformative principles, such as controlled fire or other elemental technologies that unlock the potentials of our deeper humanity. Gould and Dennett offer to us a rite of purification, which begins with a ritual of humiliation in which Copernicus, Darwin, and Freud (or other such heroes) hold out the trials, and we attempt to replicate their journeys. We also encounter something like the theme of the Tree of Knowledge: the loss of innocence as the price of painful truth (this theme will be pursued in the Conclusion of this book). Gould, Dennett, and many other science writers attribute our erroneous inclination to our inflated self-regard; but then so do many mythologies and religions.

A THINKING LESSON

Howard Margolis, a professor of public policy at University of Chicago, in his *It Started with Copernicus* (2002), carries out a highly technical analysis of the writings of Copernicus (and other cosmologists of his era) aimed in part at reasserting, against recent detractors from the field of history of science, that a Copernican Revolution did really occur. Margolis attempts to isolate and characterize just what it was that allowed Copernicus to escape the "cognitive barriers" and "habits of mind" that entrapped other thinkers of his era. Margolis aims to draw out a cognitive lesson that might be applied fruitfully in other realms of inquiry, such as his own field of public policy. While Margolis' analysis has many fascinating moments, his characterization of the distinctive quality of Copernican cognition is frustratingly vague. Margolis' project links cosmology within the history of science, on one hand, and to public policy, on the other; but Margolis summarizes the lesson he draws from the former in the language of the latter. Margolis designates the elusive Copernican cognitive quality as "Around the Corner Inquiry" ("AC Inquiry" for short) – an appellation that conjures up the fuzzy buzzword feel of contemporary "proactive" managerial-speak.

Margolis' project echoes a famous earlier invocation of Copernicus, namely that by Immanuel Kant discussing the course of metaphysics in the Preface to the second edition of the *Critique of Pure Reason*:

> The examples of mathematics and natural science, which by a single and sudden revolution have become what they now are, seem to me sufficiently remarkable to suggest our considering what may have been the essential features in the changed point of view by which they have so greatly benefited. Their success should incline us, at least by way of experiment, to imitate their procedure ... Hitherto it has been assumed that all our knowledge must conform to objects. But all attempts to extend our knowledge of objects by establishing something in regard to them *a priori*, by means of concepts, have, on this assumption,

ended in failure. We must therefore make trial whether we may
not have more success in the tasks of metaphysics, if we suppose that objects must conform to our knowledge ... We should
then be proceeding precisely on the lines of Copernicus' primary
hypothesis. (1965:21–2)

The parallel to Margolis' project, while not exact, is nevertheless evident: the underlying message is that Copernicus provided a general
model for straightening out intellectual inquiry.[8] That Copernicus
is hero may reflect a tendency to assume that cosmology, because
encompassing, is most fitting to bequeath the salutary principle for
all sub-regions of knowledge.

Kant's invocation is also interesting for another reason: it dramatizes the versatility of the Copernican metaphor. Like Copernicus,
Kant too has taken on hero-status – or for some, a sort of anti-hero
status. Bertrand Russell saw in Kant a bold but retrograde move:
"Kant spoke of himself as having effected a 'Copernican revolution,' but he would have been more accurate if he had spoken of a
'Ptolemaic counter-revolution,' since he put Man back at the centre
from which Copernicus had dethroned him" (1948:xi).[9]

Kant's use of the Copernican analogy within a work calling for
recognition of the intellect's active imposition of form on sensation
could be threatening to those who invoke Copernicus as manifesto
for the ideal of observation uncontaminated by the subject. Putting
aside the question of the validity of Russell's view of Kant, what
could be more rhetorically compelling than a deep analogy (or a
deep contrast) between a pair of modern champions who, like the
"culture heroes" of yore, reverse our course and lead us out of our
backwardness (or back into it)? The idea of the Copernican Revolution, disassociated from specifics of historical moment, becomes
applicable, as metaphor, to any change of direction that a thinker
wishes to dramatize as radical and significant.

෧෨

One other dimension of Copernicus as thinking lesson should be
mentioned, specifically, that Copernicus' finding that Earth revolves
around the sun also finds service as an epitomizing symbol of *facticity*. Consider the concluding line of Dominic Strinati's informa-
tive book, *An Introduction to Theories of Popular Culture*: "Popular

taste may be able to do nothing about the earth moving around the sun, but it can't be separated from the determination of the popularity of cultural forms" (1997:260). Although this comment contains no explicit reference to the Copernican Revolution, it alludes to a link between "popular opinion" and geocentrism, as though the latter captures the worldview or "mindset" of the former. More importantly, it offers Earth's moving around the sun to illustrate a condition we cannot influence: an epitomizing example of a *given*, a *fact*. But there is a quirk here: why should this particular fact, that Earth circles the sun – a fact that apparently is not very obvious since it eluded us for centuries – serve as the emblem of the factual? We have a number of idioms that, in ordinary speech, we draw upon as paradigmatically factual. Some of these have deductive necessity (or truth by definition) – for example, "Is the Pope Catholic?"; while others, such as Earth circling the sun, instance inductive truth (or truth gained by observation). That Earth's revolving around the sun (apparently a counterintuitive truth) stands as a privileged example of inductive facticity (vis-à-vis the inductive facticity of a thousand things that one daily stubs one's toe on) must have something to do with the *story* we have constructed around this particular fact – a story of mythic proportions, depicting the triumph of fact in a wrenching, heroic confrontation with illusion.

DOCUMENTING HUMAN INSIGNIFICANCE: THE COSMOS AS AUDIOVISUAL AID

While the theme of Copernican cosmic angst is centuries old, the versions of it focused upon so far in this chapter are datable as late nineteenth- or early twentieth-century variants – most easily by the assimilation to the Copernican Revolution of later doctrines (Darwinian evolution, psychoanalysis, Einsteinian relativism, electronic computers), but also by many features of presentation and style characteristic of contemporary science writing. But there remains another recent variation on the idea of the Copernican Revolution, one with an even more obvious twentieth-century signature. The final version of Copernican angst I will discuss does not involve explication of technical scientific analysis; it has no need of technicality because its pedagogical method is one of popular spectacle.

To be specific, one of the curious side consequences of the Apollo moon missions, mounted by NASA in the 1960s and '70s, is the stimulus they gave to a new kind of rumination about the Copernican Revolution. The revival was prompted by a technological feat: humans for the first time physically circled and then set foot on a foreign celestial body and looked back on Earth – thus experiencing the relativity of planetary position viscerally, *actually*. In terms of the Copernican Revolution, technically it would have been better if the astronauts had stood on the *sun* and looked back at Earth; yet the moon offered a convenient and, from a practical point of view, well-advised substitution – a transference facilitated, moreover, by the fact that the moon missions were named for a god nowadays popularly associated with the sun (Apollo) rather than with the moon. One could go on at length about mythological journeys to the sun (there are numerous Native American versions, for example; then there is Icarus' near miss), although other than the mission's name, there is really no evidence that the moon was other than NASA's intended target all along; this was not, in other words, an instance of the kind of internal miscommunication that a decade ago caused the crash of a Mars probe, because one technical team had calculated in English units, the other in metric.[10]

As noted above, the idea of the Copernican Revolution has often been cast as instancing or even initiating a cyclic pattern of thought within Western intellectual history: the by-now hackneyed idea is that, from Copernicus on, we have periodically undergone large-scale reversals of perspective that have something in common with the Copernican reversal, thus allowing it to serve as paradigm and perennial metaphor. Hence the prestigious line of "Copernican" thinkers (Kant, Freud, Marx, and Darwin are the most frequently nominated). But the Copernican revival at focus in the Apollo missions is not another instance of this paradigmatic metaphor. The relevant term here is not cyclicity but duality: we are here dealing with the *original* Copernican reversal – the one having to do with Earth's place in the solar system – but apprehended in a new register. The original register was abstract and mathematical, which is all we had until a few decades ago. But as of the moon missions, we have in the astronauts' small but actual steps the possibility of one giant Copernican leap for mankind. The two Copernican revolutions – Copernicus and NASA's – are, so to speak, Copernican revolutions, respectively, of the mind and of the body. The moon missions created complex *feelings* among those of us left behind to follow on television. There was a surreal palpability created by the thunder

202 The Ancient Mythology of Modern Science

of the giant rocket engines – burning ten times as much fuel in the first second as consumed in Lindbergh's entire Atlantic crossing, and leaving an initially cynical Norman Mailer repeating "oh, my God! oh, my God."[11] There was also the half-intelligible radio voices (punctuated by the familiar appeals to that ubiquitous spaceman, "Roger") and the bouncing quality of the moonwalk (too bizarre to be staged) – and our worries for members of our species off in a surreal distance.

No one was hoping that we would conclude that Earth actually orbits the moon. Rather, it was hoped that we would receive a lesson in reversibility of perspective – more broadly, in the difference between appearance and reality – by orbiting and then standing on (vicariously through the astronauts) what would *appear to be* the fixed firm moon while watching Earth rise and pass overhead.[12] The original Copernican Revolution had been based on observation, but only of the distant, patient, astronomical kind, layered in mathematical calculations. By contrast, the second Copernican revolution would be immediate, visual, and would require no math skills (only a television) – at last, a Copernican revolution for the masses!

Of the smattering of treatments of the Copernican implications of the moon missions, five stand out in creativity and elaborateness (and also in their distinctiveness vis-à-vis one another): social theorist and philosopher of science Hans Blumenberg's impressive *The Genesis of the Copernican World*; Tom Stoppard's ingenious dramatic farce, *Jumpers*; Robert Poole's detailed journalistic/historical analysis, *Earthrise: How Man First Saw the Earth*; Joel Primack and Nancy Abrams' program for remythologizing the cosmos, *The View from the Center of the Universe*; and Carl Sagan's pulpy popular work, *Pale Blue Dot*. Although the least intellectually challenging of these works, Sagan's nevertheless most fully exemplifies characteristics of the popular science writing genre, and for this reason I consider it in the most detail, commenting only briefly on the other four works.

☙❧

Carl Sagan's *Pale Blue Dot* is a manifesto of support for continued space exploration and for humans ultimately colonizing the cosmos. Given this ambition, Sagan's opening chapters might seem perverse, for they attempt to persuade us of our cosmic insignificance.

Sagan's chapter 3, "The Great Demotions," gives an overview of the developing sense of human cosmic insignificance that was initiated by the Copernican Revolution. Specifically, Sagan portrays a shrinking anthropocentric perimeter in the form of an elongated series of strategic retreats we have staged in the face of scientific advances, among which are listed the three "blows" of Freud's triadic formulations (see above). Our first attempted retrenchments involve the place of Earth in the cosmos (e.g., *"even if the Earth isn't at the centre of the Universe, the Sun is"* [1994:26]). These are followed by attempts to retrench in the face of post-Darwinian biology; *"even if we're closely related to some of the other animals, we're different – not just in degree, but in kind"* (1994:31). The final retrenchment has to do with our minds: *"God and angels aside, we're the only intelligent beings in the Universe"* (1994:32).

There is much more in *Pale Blue Dot* about Copernicus, and the great demotions theme is filled out through discussions of the projective, animistic tendencies we bring to our attempts to understand the cosmos (e.g., 32ff., 46ff., 52ff.), including the classic passage from Xenophanes on the errors induced by anthropocentrism (1994:23; see also pages 3–4). But Sagan's attempt to diminish our cosmic status ultimately does cohere with his larger purpose, for developing the motivation to colonize the cosmos requires our letting go of the notion that we are already in the one special place we are meant to be. Sagan subscribes to a sort of cosmo-existentialism: humans are an on-the-road species, and our place in the cosmos will be what we make it.[13]

But what stands out in Sagan's attempt to teach the Copernican lesson in *Pale Blue Dot* is the factor that most distinguishes his approach to the popularization of science in general, namely the glitzy visuals. Specifically, Sagan attempts to convince us of our cosmic insignificance by means of stunning photographs taken of Earth from other locations in the cosmos. He features prominently photos of Earth viewed from the moon. Among the more striking of these is a series of four time-lapse photographs that illustrate Earth progressively rising above the moon's horizon. He asks the Copernican question: "If we lived on the moon, would we consider it the centre of the Universe – with the Earth paying homage to us?" (1994:44).

Currently a number of popular books are available on the glory days of the American and Soviet space programs, many of them directed toward the same boomer-nostalgia market that has brought us books on '60s muscle cars, Harley-Davidson motorcycles, and Lionel toy trains.[14] While Sagan is in the company of many other popular writers in stressing the impact of the earthrise photos, I noticed, early on, a curious anomaly in Sagan's presentation of photos vis-à-vis those of other authors. When they take the seemingly obligatory moment to rhapsodize about the image of earthrise, most authors focus on the first such photo – the one taken by the astronauts of Apollo 8 (Figure 6.1) – and take some time to fill in the circumstances surrounding the actual taking of this photograph. Although Apollo 8 did not involve a moon landing (it was a practice-mission toward that goal), it marked the first time that humans had left Earth orbit to enter moon orbit, and thus the first time humans directly observed earthrise. And here is the anomaly: Sagan breaks with tradition by offering his readers numerous photos of Earth as seen from the moon during Apollo missions, including earthrise photos, but without once mentioning Apollo 8.[15]

A possibility comes to mind regarding this little anomaly: specifically, Sagan may have skipped mention of Apollo 8 and the details of its photographic achievements because of two related circumstances of this mission that spoiled its Copernican usefulness. The first spoiler is the date on which the mission occurred. Unless there was behind-the-scenes collusion at NASA, it was a mere cosmic coincidence (or, more specifically, a coincidence of the Newtonian-cosmic and the Christian-cosmic) that decreed that the first launch window available for the readied Apollo craft would put humans in orbit around the moon on the eve of a major feast of a religious tradition that in the past offered resistance to the Copernican cosmos and, among some adherents, still does. Specifically, the Apollo craft would go into orbit around the moon for the first time on Christmas Eve. But there was a second, related spoiler. I and many others will recall the Christmas message that was beamed to Earth by the first humans to orbit the moon: it was a reading, each member of the trinity of astronauts taking a turn, of the Biblical creation story of Genesis – the Western world's most authoritative pre-Copernican cosmological text, the nemesis text of the heliocentric universe. Following the reading, the astronauts' message concluded, "And from the crew of Apollo 8, we close with good night, good luck, a Merry Christmas, and God bless all of you – all of you on the good Earth" (see Chaikin 1998:122).[16]

In his recent, best-selling memoir, *Failure Is Not an Option*, former NASA flight director Gene Kranz, in a chapter entitled "The Christmas Story," writes: "Apollo 8 was a very big leap that drew on one's spiritual and moral resolve. For us it would become the second greatest Christmas story every [*sic*] told" (2001:238). And regarding his reaction to the Genesis reading, "I was enraptured, transported by the crew's voices, finding new meaning in the words from Genesis. For those moments, I felt the presence of creation and the Creator. Tears were on my cheeks" (2001:246). In his own recent memoir, *Flight: My Life in Mission Control*, Chris Kraft confirms that this reaction was general; he reports that there were no dry eyes in mission control nor (he was told) in the newsroom (Kraft 2002:300). For a public intellectual given to persuasion-by-spectacle, the Genesis creation story beamed from moon to Earth would be impossible to top. Sagan's title – *Pale Blue Dot* – may be a mild rebuke to the more anthropocentric media-created vision of Earth as a shiny blue marble, but Sagan surely would not have been so bold or foolish as to attempt to top the Christmas spectacle of Apollo 8, an act which ipso facto would leave Sagan as the Grinch!

Here is one recent recapitulation of the original apparition:

> For the first time, humans saw their home planet above another body: it was an earthrise. This would become the enduring image of Apollo 8, captured for their fellow humans back home by the cameras they carried on board. It was truth, light, warmth, and richness, circumscribed in a perfect circle, surrounded by an endless sea of utter blackness and shining above a dead world which had none of its luxuriance ...
>
> After Apollo 8, the world would never again see itself from the same perspective. We could no longer pretend that our world was without limits; we would see that we were an island paradise in the vast cold void of space. (Reynolds 2002:105, 110)

With similar sentiments, Andrew Chaikin, in *A Man on the Moon* describes the experience from the standpoint of astronaut Anders on board the spacecraft:

> He couldn't help but think that the cosmos would continue to turn as it always had if suddenly there were no earth. But how

little that mattered when it appeared, blue and radiant, rising beyond the lifeless moon. In that moment he saw a thing of inexplicable fragility; later he would liken it to a precious Christmas tree ornament. And if the earth was only a mote of dust in the galaxy, that blue planet was everything to him and the creatures living on it. On his way into a fitful sleep, Anders began to realize: *We came all this way to explore the moon, and the most important thing is that we discovered the earth.* (1998:119)

In *Earth Shine*, Anne Morrow Lindbergh puts it this way: "No one, it has been said, will ever look at the moon in the same way again. More significantly can one say that no one will ever look at the earth in the same way. Man had to free himself from earth to perceive both its diminutive place in a solar system and its inestimable value as a life-fostering planet. As earthmen, we may have taken another step into adulthood. We can see our parent earth with detachment, with tenderness, with some shame and pity, but at last also with love" (1969:44). Similar reactions sparked by the appearance, as though entirely unanticipated, of a delicate, luxuriant, perfect Earth, can be found with variations in dozens of books about the Apollo missions.[17] The reaction is doubly attributed: first to the astronauts of Apollo 8; second, and more insistently, to all of humanity. The view that the cosmic adventurers brought became a mythic presence: a spontaneous, powerful, collective, cosmic vision transcending any one author or moment in time and elaborated in spiralling variation around an indelible core.

Sagan's idea was that, compared with the original Copernican Revolution, a distant mathematical abstraction, this new one would take its power from the immediacy of the experience of perspectival relativity. Ironically, one can arrive at Sagan's lesson only by abstracting away the immediate experience – because, by most accounts at least, the spontaneous reaction to earthrise was not of perspectival relativity and Copernican angst, but of the splendour and specialness of a jewel against the enormity of blank, desolate space.[18] The vision inspired religious turnings and in some cases reinvigorated ancient mythological symbols; consider, for example, how James Lovelock merges his controversial "Gaia hypothesis" with the vision of earthrise: "When I first saw Gaia in my mind I felt as must an astronaut have done as he stood on the Moon, gazing back at our home, the Earth. The feeling strengthens as theory and evidence steadily confirm the thought that the Earth may be a larger state of life. Thinking of the Earth this way makes it seem, on happy days, in the right places, as if the whole planet were celebrating a sacred ceremony" (1995:192–3).

But if the moon photos did not entirely succeed in teaching the Copernican lesson, the space program offered yet another chance, this one based not on relativity of perspective, but on another kind of relativity, that of apparent size (and by implication, importance) to distance. For along with pictures of Earth from the moon (though none from Apollo 8), Sagan offers us photographic images of Earth from even more distant vantages provided by unmanned robotic space probes. Here the litany of great demotions finds its visual counterpart as Earth appears smaller and smaller, gradually assuming its proper insignificance.

Within this venture of photographing Earth from maximally distant points in space, Sagan devotes particular attention to a photograph that became possible as the Voyager probe passed beyond Neptune and Pluto.

> The *Apollo* pictures of the whole Earth conveyed to multitudes something well known to astronomers: On the scale of worlds – to say nothing of stars or galaxies – humans are inconsequential, a thin film of life on an obscure and solitary lump of rock and metal.
>
> It seemed to me that another picture of the Earth, this one taken from a hundred thousand times farther away, might help in the continuing process of revealing to ourselves our true circumstance and condition. It had been well understood by the scientists and philosophers of classical antiquity that the Earth was a mere point in a vast encompassing Cosmos, but no one had ever *seen it* as such. Here was our first chance (and perhaps also our last for decades to come). (1994:6)

Sagan describes a flurry of politics within a by-then cash-strapped NASA, which succeeded in reorienting the probe so that the proposed photos could be obtained – against NASA's argument that such a project "wasn't science" (1994:7). Sagan presents a series of photographs of Venus, Earth, Jupiter, Saturn, Uranus, and Neptune, each with the appearance of a distant and unspectacular star. Earth too, from the distant perspective of the Voyager probe, now looks like a distant dim star.

Or at least it was supposed to ... Even though there were no astronauts aboard this time to spring a reading from Genesis, this second grab for self-demotion – this cosmic act of reverse chauvinism – was a failure, too, or at best a compromised success. Unlike the other planets, the pale blue Earth dot of the Voyager photograph sits in the centre of a shaft of soft luminescence (see Figure 6.2), reminiscent of the *heavenly beacon* motif found on greeting cards of the "inspirational" variety that can be found in abundance at religious gift stores. Sagan faces a cosmic version of the embarrassment felt by any technology-age instructor when the AV equipment fails, leaving the presenter not only without the image that had been chosen precisely for its power to persuade, but with a feeling that the failure itself – the non-appearance of the announced image – will induce skepticism about the presenter's description of what the audience *would have seen*. Sagan handles the situation as well as can be expected: "Because of the reflection of sunlight off the spacecraft, the Earth seems to be sitting in a beam of light, as if there were some special significance to this small world. But it's just an accident of geometry and optics. The Sun emits its radiation equitably in all directions. Had the picture been taken a little earlier or a little later, there would have been no sunbeam highlighting the Earth" (1994:8). If the cosmic lesson in self-demotion was compromised, at least we got a cosmic confirmation of the oldest tip in documentary photography: always bracket your photos!

But the lesson in self-demotion, however flawed, is only part of the message that Sagan wants to convey. It is a near constant of science writing to offer a compensatory vision for the one that we are asked to give up. The sunlight reflected by the spacecraft, a disaster from the point of view of the Copernican lesson, can now be put to inspirational use as a positive litany, to compensate for the earlier litany of demotions. Sagan takes full advantage: "The aggregate of our joy and suffering, thousands of confident religions, ideologies, and economic doctrines, every hunter and forager, every hero and coward, every creator and destroyer of civilization, every king and peasant, every young couple in love, every mother and father, hopeful child, inventor and explorer, every teacher of morals, every corrupt politician, every 'superstar,' every 'supreme leader,' every saint and sinner in the history of our species lived there – on a mote of dust suspended in a sunbeam" (1994:8).[19]

What prompts this startling turnaround in tone from the earlier litany, the one that left us at the rock bottom of the great demotion? Sagan's is a cosmic version of the standard existentialist turnaround: once we accept ultimate meaninglessness, we are then free to create our meaning. In the cosmic context this means that once we get through the great demotions and accept our physical place in the universe for what it is, *then* we are free to wax as maudlin as we wish about our little planet and our future plans for expansion through the cosmos. And if the perspective from deep space erases our conceits about the specialness of our place in the cosmos, it compensates with something we want almost as much: a God's-eye-view of ourselves. Mirroring the spatial exchange through which, by standing on the moon, we see Earth *objectively* (as a thing in motion) for the first time, the distant cosmic source of Sagan's moral punditry lends to it a God's-eye omniscience, a kind of objectivity heretofore unavailable to mere mortals.[20]

A COMPENSATORY RE-CENTRING

Although he does not put it this way (indeed he could not put it this way because it would sound so pre-Copernican), what Sagan in effect offers to his readers, as his book progresses, is a way to restore Earth to the centre of the cosmos. Specifically, Sagan holds out the possibility of our *making* Earth into a centre by carrying out a cosmic diaspora from it. We will have to give up the idea of being at the absolute physical centre of the cosmos (though in a diasporic milieu, who cares about *that* centre anyway?), but our planet will become the emotional and physical centre – a cosmic "Old World" – to the expanding sphere of new world homesteads carved out of the cosmos by the emigrants. We will create Earth as centre by projecting our dispersal from it, whence it will loom as an absence with great presence. Sagan thus offers us a secular alternative to Dante's beatific vision, mentioned in the previous chapter: a cosmically infinitesimal point encircled by transfixed souls.

The vision that Sagan offers, moreover, is a cosmological version – a cosmological projection – of social issues very much at the centre of academic and popular interest in our era, particularly processes of diaspora, cultural retention, and cultural diversity. The many interesting but otherwise gratuitous references in *Pale Blue Dot* to Sagan's own Old World relatives and immigrant ancestors (e.g., 1994:367, 374, 402) make sense when one realizes that the immigrant/diasporic experience is the model that will hold the universe together in Sagan's grand vision. Sagan's vision involves one other important theme in contemporary society: technology. It is the power of technology, specifically the power to colonize the cosmos, that allows Sagan to advocate a radically new remedy for our anthropocentric inclinations, a remedy that involves changing our place in the universe as much as changing how we think of our place in the universe. "Our tendency has been ... to pretend that the Universe is how we wish our home would be, rather than to revise our notion of what's homey so it embraces the Universe" (1994:405).

Sagan calls for us to move out into the cosmos beyond the moon, to "terraform" other planets as habitations. As we do so there will of course be a shift in allegiance: "Our descendants, born and raised elsewhere, will naturally begin to owe primary loyalty to the worlds of their birth, whatever affection they retain for the Earth" (1994:348). Sagan manages to harmonize the fashionable topic of diaspora with the equally fashionable topic of multiculturalism, for we will cosmically instantiate both: "with human societies on several worlds, our prospects would be far more favorable ... Our eggs would be, almost literally, in many baskets. Each society would tend to be proud of the virtues of its world, its planetary engineering, its social conventions, its hereditary predispositions. Necessarily, cultural differences would be cherished and exaggerated. This diversity would serve as a tool of survival" (1994:384).

But the cosmo-colonizers will not forget their original home: "They will gaze up and strain to find the blue dot in their skies. They will love it no less for its obscurity and fragility. They will marvel at how vulnerable the repository of all our potential once was, how perilous our infancy, how humble our beginnings, how many rivers we had to cross before we found our way" (1994:405).

Colonized planets will thus be interconnected with Earth through the ancestry of the colonizers. A mythologist cannot fail to respond to this idea – in part because traditional mythologies are full of stories of migrations from now-distant homelands, and in part because the idea of the various regions of cosmos held together by ties of kinship is a quintessentially mythological idea. The example of the latter most familiar in Western traditions is the Greek mythology transmitted to us by Hesiod, detailing the genealogy of the gods that serve as tutelary deities of the various regions of the cosmos. But the notion of a kinship cosmos is also found, in intriguing variation, in mythologies worldwide.

Indeed the vision offered by Sagan recalls unmistakably the earlier-mentioned mythological imagery proffered by Copernicus in his great work, *De Revolutionibus*. Copernicus' vision, too, tenders the "heavenly beacon" motif – not, in this case, a mote in a sunbeam but rather the sun as a lamp in the middle of the cosmic temple. In Copernicus' words:

> At rest, however, in the middle of everything is the sun. For in this most beautiful temple, who would place this lamp in another or better position than that from which it can light up the whole thing at the same time? For, the sun is not inappropriately called by some people the lantern of the universe, its mind by others, and its ruler by still others. [Hermes] the Thrice Greatest labels it a visible god, and Sophocles' Electra, the all-seeing. Thus indeed, as though seated on a royal throne, the sun governs the family of planets revolving around it. Moreover, the earth is not deprived of the moon's attendance. On the contrary, as Aristotle says in a work on animals, the moon has the closest kinship with the earth. Meanwhile the earth has intercourse with the sun, and is impregnated for its yearly parturition. (1978:22)

In addition to the heavenly-beacon motif, it is important to note that in constructing and offering to their readers new visions of the cosmos to make up for the discredited "pre-Copernican" vision, both Copernicus and Sagan also fall back on an old standard: the mythico-cosmic glue of kinship. Copernicus' kinship takes the form of a patriarchal metaphor (which would not wash so well in terms of contemporary social values), whereas Sagan's is a literal, but futuristically imagined, kinship of diaspora – a sociological phenomenon that has captured the attention of our era. It is also worth noting that, as if inspired by Copernicus, the set of photographs of the planets taken by Voyager has come to be affectionately referred to (for example, in NASA web postings) as the "family portrait" (Figure 6.2). To all of the other ways in which we have accorded prototype status to Copernicus, we might add one more: he can stand as prototype for the possibility of remythologizing the cosmos in the face of scientific revolution – even a revolution as fundamental as *the Copernican Revolution.* In this respect, too, Sagan is his disciple.

Finally, although Sagan's vision projects us into the future, we should note that his concluding remarks also include gestures that place us in the centre of time, between past and future, for peopling other planets (1994:405) "binds the generations" and "returns us to our beginnings" (1994:405) – the latter thought amplified by reference to our "ancestors" of two million years ago: microbes, then fish, and so on. The genealogy goes back even further in Sagan's famous television series *Cosmos,* rivalling that of many traditional mythologies. For in the chapter "The Lives of the Stars" from that series, Sagan explains how the atoms of which we are made were produced in stars, summarizing: "We are, in the most profound sense, children of the Cosmos" (1980:198). While the line is pronounced slowly and quite movingly, in a later, more analytical frame of mind one might wonder what "most profound sense" means since, given the astronomy lesson in which it is ensconced, the phrase obviously excludes the connotation of "literally."

The foregoing points about Sagan's mythologizing of the cosmos, mainly in *Pale Blue Dot,* dovetail in a general way with one of the conclusions reached by Thomas Lessl in his study of the series *Cosmos.* Specifically, while focusing on topics other than the Copernican Revolution, Lessl nevertheless takes note of a *restorative* impulse in Sagan: "In the broadest historical spectrum, *Cosmos* represents an attempt to restore or to replace a culturally based scientific ethos that was originally rooted in Western religion, but gradually eroded as science became increasingly secularized in the last two hundred years" (1985:183). The specifics of what Lessl sees Sagan as trying to restore are, for present purposes, less important than the impulse itself – one that, given Sagan's absorption with questions about the future of our species, could easily be missed. A point that remains ambiguous in Lessl's interesting commentary is just how *deliberate* Sagan was in attempting to restore specific currents from a previous historical ethos.[21] For my reflections on Sagan's views on the Copernican Revolution, the analogous question would be: with his new way in which we would once again be the centre, did Sagan *mean to* restore elements of pre-Copernican spatial sensibility? Or, led by a showman's instincts, did Sagan simply tap into topics that he knows to move people – notably home and family – and end up inventing his own form of geocentrism? More impressed by Sagan's showman's flare than by his historical acuity, my sympathies are drawn more by the latter view.

Science does sometimes outrun myth in terms of physical description of the cosmos, but our capacity to find cathexis with the facts, with whatever is thrown at us – to re-establish the kind of visions we find in myth – shows no signs of being outrun by science. And so, thrilled by Sagan's pulpy projection we can take heart: as Earth sets below the (at long last) post-Copernican lunar horizon and we take our first "small step" into the cosmos, there is no reason we cannot look forward to a brighter day (but does the moon have days? ... a brighter day or whatever!).

DOTTY'S DOT

In addition to Sagan's work, elaborate versions of the second, lunar Copernican revolution arise in Tom Stoppard's play *Jumpers* (first produced in 1972, and revived recently in both London and New York); in the closing section of Hans Blumenberg's philosophical and sociological treatise, *The Genesis of the Copernican World*; in Robert Poole's journalistic/historical *Earthrise*; and in Primack and Abrams' *The View from the Center of the Universe*. Very different in character from Sagan and from one another, each of these works is compelling in its own way and deserving of comment.

In an intriguing twist, Stoppard, in *Jumpers*, structures his plot (such as it is; minimal to say the least) around parallel Copernican revolutions – "his" and "her" Copernican revolutions as it were. The central characters are a man of the mind and a woman of the body, sun and moon: George, a cranky and disgruntled professor of philosophy, and Dotty, an aging prima donna of stage who has a fondness for romantic harvest moon songs. George fumes and fumbles about in the world of metaphysics as he attempts to compose a rational defence of moral absolutes in a world dominated by relativism. Meanwhile, Dotty, shaken by the spectacle of the moon landing being broadcast on television, offers an embodied reflection upon it by placing a fishbowl over her head and imitating the astronaut's moonwalk in her living room. Her distress flows from the de-romanticizing of the moon brought about by humans treading on it, but also by the perspectival reversal this affords. At one point, the two

Copernican discourses – parallel but as different as day and night, betokening George and Dotty's failing relationship – align with one another:

> GEORGE: We are *all* still shaking. Copernicus cracked our confidence, and Einstein smashed it: for if one can no longer believe that a twelve-inch ruler is always a foot long, how can one be sure of relatively less certain propositions, such as that God made the Heaven and the Earth ...
>
> DOTTY (*dry, drained*): Well, it's all over now. Not only are we no longer the still centre of God's universe, we're not even uniquely graced by his footprint in man's image ... Man is on the Moon, his feet on solid ground, and he has seen us whole, all in one go, *little – local* ... (1974:75)

Although it does not argue in favour of a pre-Copernican physical cosmos, Stoppard's play does appear to affirm the wisdom of a pre-Copernican moral order. The flavour of the compensatory recentrings offered respectively by Stoppard and Sagan differ, the former subtly championing a tipping of the moral cosmos back to the former ways, the latter embracing modern existentialist angst, channelled into space exploration. Yet Sagan's futurism also returns us to our roots – to our primordial exploratory impulse – so that he, too, like Stoppard, anchors the present and future in the past. In the end Sagan invokes an even deeper past than does Stoppard, one furnished by a post-Darwinian bio-evolutionary framework, with microbes and ultimately stardust as our ancestors.

The significance that Stoppard, in the character of Dotty, derives from the moon-mission images contrasts sharply with that derived by social theorist Hans Blumenberg. Blumenberg's *The Genesis of the Copernican World* is an imposing work, impressive not only for its engagement of arguments from the history and philosophy of science, but also, perhaps especially, for its probing of the broader humanistic musings that have sprung from the idea of the Copernican Revolution. I will focus on the one point of Blumenberg's work that relates to Sagan's and Stoppard's texts. Specifically, rather than intensifying the anguish of Copernican decentring, the Apollo missions, for Blumenberg, "brought to an end the Copernican trauma of the earth's having the status of a mere point – of the annihilation of its importance by the enormity of the universe" (1987:678). Indeed, the missions were quickly followed by an unanticipated "turning of interest from the remote world to the proximate one" (1987:678), occasioned by the vision of Earth as the "cosmic oasis on which man lives – this miracle of an exception, our own blue planet in the midst of the disappointing celestial desert" (1987:685). Sagan, one will recall, invoked the image of Earth from the moon to try to affirm our Earth's cosmic insignificance, and in turn to raise support for our exploring of the rest of the cosmos. But on both of these counts, according to Blumenberg, humanity took from the images just the opposite message: we turned inward.

Blumenberg and Sagan converge, however, in crediting their respective messages to the power of photographic images. The fact that Blumenberg so credits the visual strikes home with special force, for while visual persuasion is a cornerstone of Sagan's career, Blumenberg's pronouncements on the Apollo images mark their own kind of exceptionalism, that of such visually inspired sentiments vis-à-vis the broader homage Blumenberg pays to the *logos* of the written word (nearly 800 pages in this volume alone). Blumenberg writes as one sharing, not just reporting on, the reaction of less-bookish humanity: "Language, in these events [the moon missions], remained a paltry rudiment, a mere seasoning – in spite of uninterrupted verbal production – to the stream of transmitted pictures ... If one tries to relate the centuries of imaginative effort and cosmic curiosity to the event, then the both unexpected and heart-stopping peripety of the gigantic departure from the Earth was this one thing, that in the sky above the Moon one sees the Earth" (1987:676–8).[22]

In the sources considered above, one encounters a divided opinion on the message to be taken from the photographic image of Earth seen from the moon. Sagan and Stoppard see in it a second Copernican revolution, which completes and intensifies the first by making immediate, emotional, and available to Everyman what the first time around was distant, rational, and the arcane insight of specialists. By contrast, in the general reaction to earthrise as attested in numerous popular works, and with analytical explicitness in the scholarly arguments of Hans Blumenberg, we encounter the opposite interpretation: the moon landing as easing our Copernican trauma by offering a new version of human specialness in the cosmos – emotionally, at least, a sort of Copernican counter-revolution.

Robert Poole's *Earthrise* (2008) appeared after this chapter had already been completed. It is an interesting, detailed work about the earthrise photos and their social, literary, and pop-culture impact. For one who lived through the time of the Apollo missions, the most surprising and intriguing aspect of Poole's research may be the considerable evidence he has accumulated on the behind-the-scenes politics and manoeuvrings of the space program; as is often the case, his investigation reveals the careful staging of much of what appeared at the time to be spontaneous. For example, although he discounts the view that the Christmas Eve timing of Apollo 8 was a publicity stunt (2008:133), Poole shows the Genesis reading to be the culmination of a long process of negotiation that included a major letter-writing campaign (2008:128ff.). Poole also confirms the doomed Copernican lesson from the Apollo photos; discussing debates between Sagan and others regarding the "inward" ideal of earth ecology vs. the "outward" ideal of space colonization, he says: "One name was often invoked in these discussions: that of Copernicus. The first Earth photographs were widely expected to hammer home the Copernican awareness that the Earth was not the centre of creation but just a planet among others. The problem was that in the pictures, Earth remained at the centre of the frame, appealing so eloquently direct to the viewer that the astronomical context was almost forgotten" (2008:192). Despite the impressive documentary detail and a concern with the role played by Sagan, nothing in Poole's evidence either directly confirms or disconfirms my opening conjecture concerning the curious absence in Sagan's *Pale Blue Dot* of the iconic Apollo 8 image.

The final work to discuss – and an appropriate focus around which to begin to draw this chapter and indeed this book to a close – is *The View from the Center of the Universe: Discovering Our Extraordinary Place in the Cosmos* (2006), by Joel Primack and Nancy Abrams. Rehearsing a familiar theme, these writers see Copernicus' discoveries, and their aftermath in Galileo and Newton, as abolishing the primordial mythic view of the cosmos as a home in which we occupy the central place. But they are concerned also with another change: specifically, they think that the public has fallen into the assumption that since science keeps overturning itself, we cannot expect to have again a stable vision of the cosmos. The end result of this double alienation – that we are objectively not the centre, and that we cannot have definite knowledge of the cosmos – is that we cannot connect with the cosmos in a meaningful way. They regard this as a regrettable situation and a "prejudice" that they would like to turn around. In the midst of their reflections on the cosmic alienation precipitated by the Copernican Revolution, they present earthrise as part of the remedy, accompanied by a novel rendering of the leitmotif of this photograph's singular power to bring home emotionally what we already know intellectually. "A famous photograph taken from the Apollo spacecraft by the first human beings to orbit the moon shows Earth as a sparkling blue-and-white ball, suspended in blackness, with a sterile lunar landscape in the foreground. This photo jolted many people into realizing in their hearts what of course they knew intellectually: that maps and globes have imprinted a false picture of reality on our minds. The photo showed no countries on our planet, only land masses, oceans, and clouds. The endless preoccupation with nations and racial or ethnic groups has completely misled our intuitions" (2006:8–9).

Besides the globally unifying vision of earthrise, their remedy has several other strands. One arises from their conviction that the findings of modern cosmological science resonate with some (though not all) of the ideas found in traditional mythico-religious doctrines. Seeking such convergences may be tantalizing, but it is also problematic in several ways, including that it creates an opening for developments that lead not away from national and ethnic divisions but directly into them. For to the extent that scientists are serious about re-integrating and revaluing mythico-religious teachings in terms of their success in adumbrating and providing useful metaphors for science, chauvinism stands to be transformed rather than transcended.[23] Another part of the solution Primack and Abrams propose for our cosmic alienation lies in the claim that modern cosmology reveals hitherto unrealized ways in which we are – objectively – at the centre of the cosmos: "we are central in several unexpected ways that derive directly from physics and cosmology – for example, we are in the centre of all possible sizes in the universe, we are made of the rarest material, and we are living at the midpoint of time for both the universe and the earth. These and other forms of centrality have each been a scientific discovery, not an anthropocentric way of reading the data" (2006:7).

They list, moreover, six ways in which Earth is "an unusually suitable planet for life" (2006:210ff). In *The New Universe and the Human Future* (2011), based on the Terry Lectures that they gave at Yale University in 2009, Abrams and Primack repeat the litany of six just as "a seventh way that Earth is highly unusual may have been discovered" (Abrams and Primack 2011:146). It would seem that the Seven Days (of creation) are now superseded by the Seven Ways (in which Earth is special).[24] Will the new sense of Earth exceptionalism satisfy our anthropocentric appetites as well as the old mythic one did? This new program of remythologizing is as problematic as it is fascinating; it will require a separate treatment.[25]

In the concluding chapter of *The View from the Center of the Universe*, Primack and Abrams contrast the view they offer, which they call the "meaningful view," with the one they regard as more commonly promulgated by scientists, the "existential view" rooted in a heroic acceptance of our cosmic insignificance. They portray Sagan as epitomizing the existentialist view (2006:274ff.). Now, this claim about Sagan is clearly warranted, but so is the reverse: no one has done more than Sagan to conjure from science the possibility of a "meaningful" relationship with the cosmos. Primack and Abrams have incompletely read Sagan's existentialism, failing to note the reconstructive part of the cycle, the rebuilding of meaning that follows the bottoming-out. In the end, the messages of Sagan and of Primack and Abrams are very similar: in both is an insistence that we give credence to the hard teachings of science as well as a conviction that new forms of emotional energy and connection to the cosmos are to be had once we do so. Moreover, in both cases the reinvigorated cosmic cathexis will arise from reinterpreting and redeploying images familiar in ancient mythologies. The substance is thus similar; the difference is that Primack and Abrams (E.O. Wilson and a few others occasionally voice similar sentiments) are openly, robustly, and programmatically calling for what Sagan (and, to some degree, most popular science writers) have all along been doing subtly if not surreptitiously.

Arguments taken to the extreme have a way of turning against themselves; and since cosmology by definition deals with extremes (the first cause, the most distant point in space, and so on) such turnarounds occur with regularity. Two instances have already been noted. Chapter 2 called attention to the ease with which arguments can be inverted within the classic "chain of being" model of the cosmos. Chapter 3 discussed a biological example: specifically, how Gould's insistence that humans are not a predetermined end point but rather an unlikely accident within evolution, ends up fueling the belief that we must have been intended after all (since we did occur even though we are unlikely).[26] A different kind of turnaround can be seen in the conclusion of Steven Weinberg's *The First Three Minutes*, with which this chapter opened. Weinberg is a renowned scientific cosmologist with a reputation for defending a pure vision of science – one component of which is a kind of austerity, a renunciation of the search in science for cosmic solace. One of the most repeated lines in popular science writing – a summarizing comment that has taken on a life of its own – occurs in the concluding section of *The First Three Minutes*: "The more the universe seems comprehensible, the more it also seems pointless" (1984:144). But consider also the momentary warming of the heart that accompanies this sentiment:

> However all these problems may be resolved, and whichever cosmological model proves correct, there is not much of comfort in any of this. It is almost irresistible for humans to believe that we have some special relation to the universe, that human life is not just a more-or-less farcical outcome of a chain of accidents reaching back to the first three minutes, but that we were somehow built in from the beginning. As I write this I happen to be in an airplane at 30,000 feet, flying over Wyoming en route home from San Francisco to Boston. Below, the earth looks very soft and comfortable – fluffy clouds here and there, snow turning pink as the sun sets, roads stretching straight across the country from one town to another. It is very hard to realize that this all is just a tiny part of an overwhelmingly hostile universe. (1984:143–4)[27]

In popularizations of science the persona of the author is crucial. The non-specialist must entrust his/her cosmic journey to a guide – rather like the cosmic wayfarer in Dante's *Divine Comedy*. With his austere vision of science, the persona of Weinberg might seem rather unpromising material for visual media, but his part in a recent video documentary on science and religion (*Faith & Reason*) reminds us just how varied the forms of personal magnetism can be. Particularly captivating is a comment in which, after affirming his vision of a cosmos indifferent to us, Weinberg says, hesitatingly, almost reluctantly, "although we are not the stars in a cosmic drama, if the only drama that we're starring in is one that we're making up as we go along, it's not entirely ignoble that faced with this unloving, impersonal universe we make a little island of warmth and love, and science and art, for ourselves. That's not entirely a despicable role for us to play" (in Wertheim 1998; my transcription). There is nothing so moving as when an eminent sage, practiced in delivering an austere message, waxes poetic in a fumbling, maudlin, but sincere way. Weinberg emerges as a microcosm of the vision of the cosmos he presents, offering a reluctant but arresting moment of warmth. Even in his purist commitment to post-Copernican austerity, he manages to contribute to one possible replacement for pre-Copernican contentment: an image of Earth as a sort of a warm, cozy parlour in the midst of a winter storm. His island vision distantly recalls, moreover, the Greek origins of the very idea of *cosmos*, which Hesiod portrays as a state of humanly habitable order emerging from the abyss of *chaos*.

രൗ

I will close by mentioning a chance occurrence that befell me after this chapter was otherwise complete. On visiting, for unrelated reasons, the government documents section of the Indiana University main library, I happened to notice NASA's own publication on the photography of the Apollo 8 mission (*Analysis of Apollo 8 Photography and Visual Observations*), which had been chosen for an exhibit of the sorts of documents contained within this depository. The introduction, by Richard Allenby, describes, as a prime scientific task of the mission, photography of the lunar surface; this would be valuable for "approach topography and landmarks for the early Apollo landings" and toward knowledge of the "broad structure and characteristics of the lunar surface" (1969:vii). Thereupon follow some 300 pages of the most excruciatingly dry, precise, technical data on photographic contrast, colouration, angle, and so on. Given these goals, can anyone guess what Apollo 8 photograph appears on the cover of this publication about photographing the moon? Hint: except peripherally, it's not of the moon. Another hint: the librarian, on seeing the document I was checking out, nodded toward it and volunteered, "It's one of my favourite photos." A cosmic version of thumbing through vacation snapshots: we naturally look for ourselves first.

Buoyed by the Apollo images of a dull, dead moon framing a lush blue Earth; by fluffy clouds over Wyoming; by arguments about the intricacy of the conditions necessary to support life as we know it and the non-discovery as yet of other life in the cosmos; by a litany of ways in which Earth is an unusually habitable planet; by support drawn from multiverse theory that we are "in a sense the lords of creation";[28] by the peculiar twist in which ideas cosmological turn against themselves, allowing Copernican austerity to metamorphose into geocentric euphoria; and all of these magnified, for some at least, by a desire for a compensatory vision to fill the void created by our Copernican decentring, the idea of Earth exceptionalism seems to be on the rise. It is an idea to watch.

Paradise Lost, Paradise Regained: How Scientists Save Myth

Apollo racing across the sky in his fiery chariot, making heavenly music on his golden ukelele – oh, that was for me, I loved it! And then along came another Greek to spoil it all, a Greek who would rather think than worship sun gods. Anaxagoras was his name.

Mr Sun, in Our Mister Sun[1]

And then along comes Darwin and spoils the picnic.

Daniel Dennett, Darwin's Dangerous Idea[2]

Then science came along and taught us that we are not the measure of all things, that there are wonders unimagined, that the Universe is not obliged to conform to what we consider comfortable or plausible ... This is surely a rite of passage, a step towards maturity. It contrasts starkly with the childishness and narcissism of our pre-Copernican notions.

Carl Sagan, Pale Blue Dot[3]

It is no secret that the Biblical story of Adam and Eve ends up at loggerheads with the scientific view of human origins; we encountered this clash several times in the foregoing chapters, most specifically in Gould's referring to the Genesis account as "mythology" that is "not an option for thinking people" (1997:19). But just because we are accustomed to the Biblical Genesis account appearing as the opposite, the nemesis, of scientific cosmology, the point is easily lost that on another level the theme of paradise lost is tapped repeatedly by popular science writers as a framing device for their expositions and recommendations. Some of their portrayals of our original pre-scientific state converge with the Biblical idea of original sin, for the root problem in both cases is seen as our proclivity to be led astray by inflated self-regard.

As typified in the epigraphs above, such portrayals hint at an untroubled, even idyllic, initial state. They provide a frame for reflection about the desirability of that state: was it unambiguously good? and is the post-fall state unambiguously bad? The theme of paradise lost, in other words, carries with it the captivating ambiguity of the "happy fault" (*felix culpa*) of Judeo-Christian eschatology. On one side lies bliss, but an ignorant bliss; on the other a life of toil, but also the Tree of Knowledge and the possibility of paradise regained.[4]

COMPENSATORY VISIONS

In urging us to foreswear the vision of ourselves as preordained pinnacle of the natural world, the great challenge to the popular science writer is to construct an attractive alternative – a compensatory vision. The idea of compensatory visions is adopted here from anthropologist Ruth Benedict,[5] an early student of Franz Boas who, like many of her generation, was also strongly influenced by Freud. Benedict took note of the fact that myths sometimes offer not accurate but, rather, distorted renditions of the societies in which they exist. Benedict hypothesized that myth, by offering fantasy, helps make up for onerous and difficult realities of life: one of the functions of myth is precisely to "compensate" for the demands of living in society. The implication is that success in offering a compensatory vision is part of what gives myth its power and appeal.

Popular science writing has an analogous function, for it is a near constant of this genre to offer a compensatory vision for the one we are asked to give up. But notice that this formulation alters Benedict's. Benedict says that myth offers fantasy to compensate for experienced reality. The challenge perceived by popular science writers, by contrast, is to show that reality as envisioned by science compensates for the loss of the fantasies allegedly offered by myth – although this is often accomplished, I argue, through the construction of a new myth. In various forms and degrees of explicitness, some sort of "trade-in offer" is a recurrent theme in this genre.[6]

FIRST THE BAD NEWS, THEN THE GOOD

Through scientific understanding, we suffer loss of innocence, cosmic dethroning, and a variety of specific hard truths that challenge cherished preconceptions and shatter our sense of cosmic specialness and meaning. What do we get in return? The full answer to this question is complex and subtle, as popular science writers typically formulate the terms of the trade by implication only. Among specific forms of compensation, the most recurrent are the following.

Truth

To give us truth, or at least a theory superior to pre-scientific understanding, is the basic and obvious commitment of popular science writing. I do not doubt that, for most readers, the promise of truth holds a sui generis appeal. But many popular science writers attempt to offer – and their success depends on – more than a better theory about specific phenomena of nature.

Maturity

One of the ways in which a hard truth can be made attractive is through the claim that the person who accepts it is superior to the one who does not. A particularly potent form of superiority, one we all experience intimately, inheres in the hierarchical relationship between adult and child. This hierarchy exudes its own *felix culpa*, since, even with their greater knowledge and powers of control, most adults experience at least moments of nostalgia for the world of the child. Despite these moments, however, and despite the charm of some of those who refuse, most people do want to grow up, even if only for their children. Tapping into this desire, popular science writers frequently ask their readers, in effect, whether they would like to grow up. In Chapter 5 we considered Daniel Dennett's juxtaposing of the hard truths of Darwinism to a Sunday-school song whose innocent vision is "a myth of childhood" that "most of us have outgrown" (1996:18); he thus superimposes on a contrast between science and traditional religious belief a contrast between adult and child. While particularly striking, Dennett's is only one of many instances of this strategy to be found in popular science writing.

Some popular presentations of science play on another form of maturity, that of the saint or sage – the individual who has arrived at a higher existence in this fallen world. The dedication of such figures is typically signalled by a disconnection from physical need and social convention. Such characteristics reappear in the heroizing of great scientists. Consider, for example, this description of Stephen Hawking by biographer John Boslough:

> As I came to know Hawking, the truth became apparent. His accomplishment is not due simply to his will to live or to the fact that he is a survivor, though he is certainly a tough and stubborn man. He succeeds because of his intellect, and as the ravages of his disease have, over two decades, taken his physical powers from him, he has come to live a life of the mind.
>
> Hawking's mind is his most powerful tool. It is also his work, his plaything, his recreation, his joy – his life. His wheelchair gives him a special vantage point for the major preoccupation of that mind: the universe we inhabit, how it came into being, how it operates, and how it will end. A totally cerebral man, he demonstrates the power of the human intellect to fathom the universe when the restless mind is set free. (1989:3)

The passage implies the dualistic worldview of religious asceticism. Body and spirit are locked in a zero-sum contest: the more one rids oneself of body, the more one releases the power of spirit, or, in the case of Hawking, of pure mind.[7] Hawking's is a particularly striking case, but less dramatic accounts, favouring the theme of unkemptness in dress or other manifestations of obliviousness to physical circumstances, flourish in the heroizing of eminent scientists. Intimations of "spirit" and "pure mind" swirl around such figures despite the fact that their craft has generated considerable skepticism toward such concepts.

As discussed in chapter 6, modern existentialist thought has carried over elements of the schema of religious asceticism: barrenness of meaning – cosmic absurdity – becomes a springboard for a heroic spirituality. Influenced by this moral schema, Sagan offers an offbeat (or perhaps "beat") cosmic pop-existentialism blended with lustre and glitz designed for those possessing meagre powers of gratification postponement. The human species is on the road, ready to take hard knocks in exchange for the adventure of self-creation – and in exchange for visiting more of those awesome stars and planets.

A Cosmos of Wonder

Gould makes a pointed comparison that describes the compensation he offers: "for sheer excitement, evolution beats any myth of human origins by light years" (2003:218). The comment encapsulates the relationship Gould sees between science and myth: they are sparring partners, the competition in this instance revolving not around which is more truthful (for Gould, that assuredly is science) but which is more captivating.

It seems to be part of our human constitution (or perhaps, to some degree, of living things in general) to be intrigued by and curious about our habitat and the other kinds of beings that reside there. One of the ways in which science, like mythology, can engage human curiosity is by offering hitherto unknown objects – indeed new worlds – for consideration. But the power of mythologies to fill unknown regions of the cosmos with imaginative creations, brilliant as it is, is matched by the instruments of science: a mere change of power of magnification in a telescope or microscope brings into view mysterious new entities and regions, seemingly without end.[8] Alongside astronomical photographs, Sagan presents a number of works of space art, i.e., artists' renderings of possibly hospitable foreign planets and imagined settlements. These offer enticing panoramas of surreal mountainscapes and iridescent colours that are both futuristic and "retro," beckoning to our cosmically diasporic destiny even while recalling the low-budget painted backdrops of mid-twentieth-century science fiction films. Need I say it? The cosmos offered to the reader by popular science writers frequently borders on enchantment.

But an enchanted cosmos requires more than an assemblage of enchanted objects; there also must be a sense of overall coherence. Integrative and/or totalizing syntheses radiate power and appeal, whether proffered through the symbolism of science or that of mythology; for academics, this appeal dovetails with a wave of interest in interdisciplinary perspectives, giving some works of popular science writing resonance within the academic community. In the foregoing chapters we have encountered many attempts by popular science writers to show readers not just interesting collections of natural phenomena, but collections coherently and enticingly arrayed. Most obvious are Barrow's chart of the ribbon of possibility running from atoms to galaxies and Gould's curve of distribution, running from the left to the right wall of LIFE. Each chart in its own way offers a *spectacle*, a synoptic view (arranged from smallest to largest beings) which, while selective, nevertheless implies a place for everything. A powerful symbol of quantitative analysis – for Barrow, intersecting vectors; for Gould, a statistical curve – holds the spectacle together, asserting, amidst an insistence on a general condition of contingency, the discovery of an underlying principle of unity; and the whole exudes the authority of science.

While the appeal of totalizing visions is most obvious in such graphic spectacles, we should not ignore the subtler examples. The power of totality is also present in Lakoff and Johnson (as in their precursors, Durkheim and Mauss) and in the project of AI homunculism; in these instances it takes the form of an encompassing schema defining the nature of human knowledge. And under the swashbuckling surface, the principle that will hold Sagan's cosmos together is also rather elementary: the nomadic propensity of our own species, which he appeals to us to recognize and reaffirm. "The Cosmos extends, for all practical purposes, forever. After a brief sedentary hiatus, we are resuming our ancient nomadic way of life. Our remote descendants, safely arrayed on many worlds through the Solar System and beyond, will be unified by their common heritage" (1994:405).

Continuity of Religious and Humanistic Values

One encounters recurrently in popular science writing a defensive but committed and hopeful voice: the old moral values will not die, but in updated form will take on new authority. Art will not succumb; rather its interests will find common ground with those of science. The old wounds of divided selves, the dichotomy of rationality and feeling, and the abyss of the "two cultures" will be healed. For those in the humanities, science will make available new source material: inspiring ideas, metaphors, and images, all carrying with them the special authority of science. Stunning visual imagery – for the cosmos is the ultimate image bank – and verbal artistry that incorporates graceful writing and personal experience will serve as vehicles of scientific exposition and even of scientific reasoning.[9]

In the foregoing chapters I have argued that many of the strategies employed in pursuit of such synthesizing ambitions – including storytelling, heroizing, speculative origin scenarios, microcosm/macrocosm analogies, bold celestial imagery, and reading of moral lessons in the structure of the cosmos – are reminiscent of the strategies characteristic of traditional mythologies. It is the poetic pronouncements about the meaning of humanity – Weinberg's comment about the seeming pointlessness of the cosmos, or Hawking's about knowing the mind of God – that become the gems of this genre, not technical theories or mathematical formulas. Of course, these are precisely what many readers are looking for, and even some of the more scientifically tough-minded stand ready to forgive poetic indiscretions if overall they benefit the cause of science.

Yet this new mythologizing is precarious, for it risks awakening the criticisms that have been levelled against the same strategies when they occur in archaic or traditional mythologies – criticisms that can be summarized as anthropocentric bias. The cautionary tale most often cited by popular science writers themselves is that of the scientifically stultifying consequences of the pre-Copernican, religious worldview; ironically, such writers play into the engendering conditions of such stultification by entangling their findings in particular moral lessons and visions. Even though subtle, the examples of such strategies that I have selected for this book are admittedly among the easier targets, or at least objects of fascination, for the mythologist's skeptical eye; yet these strategies are characteristic, in varying degrees, of much, if not most, popular science writing.

There is great variation in the ways that popular science writers deal with religion; they sidestep, wax vague or ethereal, use innuendo, claim non-interference or even broad convergence of scientific findings with religious and/or scriptural ideas, or occasionally blatantly call for the abandonment of certain religious ideas or even religion in general. Often they present science as offering immanent grounding for the basic human values and moral principles that religious traditions claim to derive from transcendent sources. Recall, for example, the arguments of Lakoff and Johnson, and their precursor Durkheim, considered in chapter 4. Durkheim thought religious symbols were concretizations of the moral forces generated by human collectivities, played out in the context of ritual. His enthusiasm for society as a final metaphysical reality is evident: it seems not to have occurred to him that some believers would balk at the loss of a transcendent divinity once this idea was revealed to be the illusion cloaking the very real experience of society's collective force. Perhaps Durkheim thought that this new immanent source of morality – society – had enough appeal to balance the loss of divine transcendence. The biggest attraction for Durkheim lay in the authority of science, for he thought that unlike transcendent divinities, the social collective was scientifically observable and measurable.[10]

One finds parallels to Durkheim in the contemporary metaphor theorists Lakoff and Johnson. They too offer, in the bodily experience of kinship, an immanent source of morality (of course it does not hurt their case that this immanent source, the body, is already an object of worship in their society). Again like Durkheim, Lakoff and Johnson link their case for an immanent source of morality to claims of an immanent source of reason (in contrast to a doctrine of "pure reason"). Through all of the specifits runs a recurrent theme: what religion and metaphysics in the past offered as claims about divinely-transcendent properties, science can now put forward in the form of claims about immanent, observable properties. Something a little different yet parallel is implicit in Barrow's fire myth: we are asked to curtail our fantasies about human cultural and aesthetic plasticity in favour of an enhanced recognition of uniformity rooted in the constants of nature; in return, these constants, revealed by physical and mathematical science, will be our cosmic place holders – in effect guaranteeing the special significance that hitherto has found its main support in religious worldviews and traditional origin mythology.

Cosmic Kinship

In researching this book, one surprise has stood out above all others: the frequency and versatility of images of kinship in popular science writing. That this is surprising may be related to the fact that, unlike the other compensations summarized above, kinship is not routinely named by popular science writers as a boon we can expect from science; they prefer to draw kinship in through more subtle ploys. In a cultural milieu given to individualism, perhaps to overtly acknowledge a desire for kinship would be embarrassing, or perhaps the desire is seen as too obviously mythogenic.

An indirect route by which kinship finds entree is through Darwinian theory, which imbues and inspires contemporary popular science speculation not just in biology, cognition, and the origin and growth of culture and aesthetics, but even the evolution of the universe or multiverse – as evidenced, for example, in analyses that approach the question of which universes, stars, or planets will occur, and/or which will ever become known (in turn requiring a planet to be a suitable habitat for the evolution of a knower), as though such cosmic processes are analogous to those of natural selection.

Observations made of living things are, in effect, projected onto non-living things. The cosmos portrayed in traditional mythologies is typically a cosmos held together by kinship; but kinship images have also appeared repeatedly in the nominally scientific expositions we have considered in this book: in Copernicus' image of a family of planets surrounding the sun as ruler; in Sagan's emphasis on the cosmic-diasporic future of our species as well as our deep microbial and even stellar ancestry; and in the notion of kinship as the imminent source of morality and even politics in the cognitive research of Lakoff and Johnson as well as their precursors Durkheim and Mauss, for whom it ends up also, in a sense, inside every classificatory category, ready to imprint its form on all human knowledge. Other important conceptualizations that we have encountered are but "once removed" from kinship: the idea of the category has not yet freed itself from the language and imagery of kinship; and the "mythology" of AI homunculism is built on metaphors of bureaucracy, the modern-world's immediate successor to social organization based on kinship. The title of Gould's posthumous popular book, *I Have Landed* – suggesting at first glance a bit of self-parody – turns out to be a phrase taken from his grandfather's letter of arrival on immigrating to America (Gould 2003:1). Like Sagan, Gould makes much of his immigrant ancestry in portraying his own moral and professional mainsprings; this contextualizing of science within the kinship of its practitioners serves in its own way to humanize the cosmos of science.[11]

⌖

Beyond attempts to offer a source of morality in the abstract, popular science writers sometimes try to read specific moral lessons in the structure of the cosmos. One of the great moral debates of the twentieth century, now carrying over into the twenty-first, is the controversy over how much of human behaviour should be attributed to nature as opposed to nurture. In chapter 2, we saw Barrow using the principles of constraint that influence all matter to cosmically ground his argument in favour of greater recognition of the nature pole. Gould, too, attempts to offer a cosmic morality lesson, in his case drawn from evolutionary theory. Gould's lesson turns out to be a variant of the oldest moral lesson in the world, a lesson that was mythico-religious before it was scientific: the need to curb human pride. Though the theme of pride – specifically, what counts as prideful – shows great variation across its many mythological, religious, and scientific inflections, I can only conclude that the need to subdue arrogance must be THE basic morality lesson, one that humans *want to hear*, preferably from high authority.

Gould is only one of the many popular science writers who embrace Copernican decentring as hallowed symbol of, and rite of passage to, the scientific attitude. However, when ideas are taken to extremes – as they necessarily are in cosmology – they sometimes give rise to their opposites. Of the several instances of this process considered in the previous chapters, the most intriguing arises in Copernican decentring: for the more insignificant Earth is, the more significant it is. The idea of Earth as a tiny oasis of life in a vast, indifferent cosmos, especially when accompanied by stunning photographs like the earthrise images captured in the Apollo moon missions, cannot but stir feelings of the delicate and the rare, and perhaps reawaken old mythic imagery (recall from chapter 6 the synergy between the earthrise photographs and the so-called Gaia hypothesis). The trauma of post-Copernican decentring thus finds compensation in the possibility of re-envisioning Earth as a fragile cosmic exception – a vision that, even if less obviously anthropocentric than the discredited vision of a geocentric universe, at least does not leave cosmic solace-seekers entirely bereft.

PROMETHEUS UNCHAINED EXPLAINED

The mythic figure of Prometheus transcends the divide of "the two cultures" by appearing repeatedly as a favoured symbol in both art and science. Scientific allusions to this figure range from terse references to "Promethean" minds and feats in the history of science to the elaborate, updated account of the origin of culture through fire that we considered in chapter 2. As much as the technology itself, Prometheus has come to stand for the boldness that inspired the theft, and gift, of fire. Prometheus conveys a defiance of oppression and fatalism, a hope for the self-directedness and cosmic maturity of the human species. Similarly, in the works considered in the foregoing chapters, emphasis is accorded less to the technologies that science has brought us than to the insights science and its technologies offer about our nature and our place in the cosmos.

If relatively few works of popular science writing have dazzling technological innovation at their core, it may be because by now we have seen the downside of many forms of technology. Or, more likely, the situation reflects an acceptance of the everyday reality of technological advance; the more pressing need is for someone to tell us what science means and what we as humans mean in a world informed and transformed by it. Freed from the shackles and inescapable destiny imposed by myth, we nevertheless want to regain one part of what we think myth offered in the past: a coherent cosmic vision that answers questions that we cannot help asking: Who are we? How did we get here? Why? What is to be done? In trying to parlay scientific findings into compelling visions and wisdom for life, popular science writers merge into an eternal return, offering new variations on what humans have always come up with – new variations shaped, if more subtly, by the same anthropocentric slant that, from the presocratic philosophers to the present, has been held up as myth's intractable flaw. I have made numerous specific criticisms of the writings I have considered, but my overarching criticism is that popular science writers, typically in the quest for polemic, persistently oversimplify and misrepresent their relation to myth. Why do they continue to mythologize? No doubt for many reasons – among them, to gain celebrity, to communicate earnest convictions and promote science, and sometimes to offer a sop to the masses. But, as well, perhaps they do so because there are some things that myth still does best and for which as yet we have found no alternative.[12] Until we know the "mind of God," we do well to keep in mind the verdict of the Enlightenment: mythology is begotten by ignorance, or to put it less harshly, by thinkers whose reach exceeds their grasp. If I am right, it is not merely in the questions popular science writers ask, but as well in the strategies they employ to answer them, that the spirit of mythology, stridently but superficially chastened, continues to flourish.

Notes

PREFACE

1   Sentiments felt as deeply personal are often widely, or even universally, shared. By coincidence, shortly after I wrote this reminiscence, one of the survivors of the ill-fated flight broke a long-standing silence by writing a book about the experience (Nando Parrado, *Miracle in the Andes: 72 Days on the Mountain and My Long Trek Home* [2006]). I found my uncle's sentiments paralleled in the words of this survivor, who, instead of "just matter," uses the phrase "only meat": "I decided to speak. 'We must believe it is only meat now,' I told them [the other survivors]" (2006:97). On the following page the author adds, "I understood the magnitude of the taboo we had just broken" (2006:98).

INTRODUCTION

1   There are relatively few exceptions to this usage; the most interesting is found in E.O. Wilson (see chapter 1, page 21).

2   While the "ethnic diversity" cited in this passage might seem to imply the issue of ethnocentrism, the larger context suggests that it is anthropocentrism – the human species in relation to gods rather than human ethnicities in relation to each other – that ultimately interests Xenophanes. Further comments on the famous passage are found in chapter 1.

3   There are many varieties of skepticism. As an observer not only of popular science writing but as well of contemporary skeptic/rationalist organizations, I cannot fail to notice the attraction that these two movements hold for one another. They also share many ideas and strategies, including a proclivity to invoke myth as a foil. The popular science writers on whom I focus my particular brand of mythologist's skepticism are regarded quite

differently in the skeptic/rationalist movement: as evidenced in conferen-
ces and such magazines as *Skeptic* and *Skeptical Inquirer*, eminent popular
science writers are greeted by skeptics/rationalists as none other than sages
and heroes.

4   For example, see Turney (2001:240); Jurdant (1993:367); Lessl (1985
    passim.).

5   Central to the "science wars" are Paul Gross' *Higher Superstition: The
    Academic Left and its Quarrels with Science* (1994) and the notorious
    "Sokal's Hoax" episode (see, for example, Sokal's retrospective on the epi-
    sode [1998], and the analysis, from the perspective of the history of sci-
    ence hoaxes, by Walsh [2006:227ff.]). The "hoax" refers to an incident in
    which physicist Alan Sokal submitted a paper laden with scientific errors
    to the cultural studies journal *Social Text*; Sokal used the fact that the
    paper was accepted for publication to dramatize what he saw as fraudu-
    lent attitudes toward science characteristic of some postmodern discourse.
    I agree with Sokal on this point, and see his perpetrating of the hoax as
    largely justified. However, something of the reverse situation is also pos-
    sible; indeed this book is motivated in part by what I see as frequently
    inadequate handling of cultural factors, especially the concept of myth and
    related concepts, by popular science writers. None of this is to discount
    the goal of science/humanities synthesis, but only to affirm its elusiveness.

6   "Comparative mythology" is often used in a narrow sense to designate the
    study of relationships between mythologies within the Indo-European lan-
    guage family. I use the term in the broader sense to designate the compara-
    tive study of mythologies wherever they occur.

7   Lakoff and Johnson's approach shares with the other writers I consider a
    commitment to scientific method (including experimental protocols) and
    to materialist explanation (i.e., to neural structures over ethereal soul sub-
    stance); an affinity for mechanisms that can be related to modern bio-
    evolutionary theory; an interest in discipline-transcending synthesis; an
    accessible style with non-technical, everyday illustrations; and a humanis-
    tic commitment to explaining how their theories relate to individual and
    social well-being – and thus why such theories should be matters of con-
    cern to everyone.

8   The Copernican Revolution, our current metaphor of choice for the prob-
    lem of anthropocentrism and the possibility of its transcendence, also con-
    notes the large – the cosmic. Thus this metaphor, too, may work against
    recognition of "centric" strategies that operate through small, subtle, non-
    spatial moves.
        Indeed it seems that with regard to many forms of epistemological and/
    or moral prejudice, we take aim at and eventually surmount the largest,

most glaring instance only to find that we still confront more subtle versions of it – or, worse yet, that we have created new versions of the same bias with our remedies. Such experiences raise doubts about how deeply we understood the bias in the first place.

9 The Gaia hypothesis provokes intense and varied response – in the scientific community, popular culture, and various ecology and spirituality movements. For a history of the early phase of the debates see Joseph (1990).

10 For similar reasons I deal only in passing with the works of those few popular science writers who explicitly attempt to interrelate science with religious/theological arguments or traditions (e.g. Paul Davies, Fritjof Capra, and recently Francis Collins in *The Language of God: A Scientist Presents Evidence for Belief* [2006]).

11 Hybridism, as a metaphor in contemporary socio-cultural analysis, arises from transposing a scientific concept to the realm of culture and society – reversing the direction of most of the metaphors that I consider in this study (including the cosmos-as-city example). Bruno Latour (1993) places the image of hybridism at the centre of a broad commentary on the mixed status (i.e., at once scientific, social-scientific, political, religious, etc.) of much contemporary academic and journalistic discourse. My own interest in this metaphor, however, is inspired more by detailed analyses of processes of blending between specific folkloric and/or literary genres (see, for example, the special issue, "Theorizing the Hybrid," of the *Journal of American Folklore* edited by Kapchan and Strong [1999]; cf. Bauman and Briggs [2003]).

12 The phrase, a slogan of the Ringling Bros. and Barnum & Bailey circus, was adopted recently by Richard Dawkins as a book title (*The Greatest Show on Earth: The Evidence for Evolution* [2009]).

13 Lucretius was a gifted poet; many of the artistic devices and strategies he employed in his treatise on the cosmos, *De Rerum Natura*, to wean his readers from traditional religious beliefs and, in compensation, to render the world of atoms exciting and enticing are to be encountered with only slight modification in contemporary popular science writers (Schrempp ms. in progress [a]).

CHAPTER ONE

1 While Snow's assessment of the relationship of scientific and literary culture contains many asymmetries that seem to favour science, these do not lead him to the sort of usurping vision that is suggested by Brockman (and to some extent also by Wilson: see note 5, below). Snow's 1959 lecture

as well as his "second look" can be found in Snow (1998). Many other scholars have weighed in on the "two cultures" debate (e.g., Slade and Lee 1990; Shaffer 1998; Labinger and Collins 2001). Also see Ben Shneiderman's (2003:2) discussion of the debate in the opening of his *Leonardo's Laptop*, a popular work about the challenge of humanizing innovations in computer and Internet technology.

2   Brockman's fuller commentary on his and Snow's usage is found in *The Third Culture* (Brockman 1996:17–19).

3   Smolin's *The Life of the Cosmos* (1997), with which I opened my Preface, offers a good example of such interdisciplinary reach. Specifically, Smolin brings Darwinian theory to the evolution of stars and planets. Even more impressive is his engagement of the tradition of philosophical cosmology, especially the arguments of Leibniz.

4   Ongoing discussions of these integrative concerns can be found in a website launched by Brockman: Edge (www.edge.org).

5   As one proceeds in Wilson's book, it quickly becomes evident, however, that Wilson's "consilience," like Brockman's "third culture," carries hegemonic overtones: any synthesis that emerges, it seems, will do so from the side of science. See also Eger's (1993:190) commentary on Wilson and the theme of bridging art and science.

Among the many other endorsements of the desirability of such bridging, Steven Pinker's Introduction to *The Best American Science and Nature Writing 2004* also stands out. Noting that as editor of this volume he looked for essays that combined "explanatory depth" with "limpid prose" (2004:xiv), Pinker opens with the comment: "Horace's summary of the purpose of literature, 'to delight and instruct,' is also not a bad summary of the purpose of science and nature writing" (2004:xiii).

6   A variety of science narratives will be touched on in the course of my analysis. Additionally see Dawkins' (2005) expounding of biological evolution through the format of Chaucer's *Canterbury Tales*, the recent collection of scientists' autobiographical narratives compiled by John Brockman (2005), and comments about storytelling by Alan Lightman (2005) in his introduction to the anthology *The Best American Science Writing 2005*.

7   Indeed, it seems that there is a tendency to imbue any story, image, or idea that looms large in consciousness with at least a superficial aura of the primordial and numinous. See the comments below on Roland Barthes' (1995) discussion of ancient and sacred allusions in seemingly modern, secular mythologies.

8   Reflections on the complex and historically shifting valences of true and false, and on the epistemological status of myth in general, are found

throughout Veyne (1988), Detienne (1986), Lincoln (1999), and Brisson (2004). Drummond offers an elaborate excursus on myth in the modern world, emphasizing our "profound ambivalence toward the role of myth in our lives" (1996:25, passim.).

9   For a discussion of the tension between narratives that are rigorously accurate vs. emotionally moving in the exposition of science, see Alan Gross' analysis of James Watson's controversial classic, *The Double Helix*. Gross builds his analysis around Watson's "choice of psychological over literal truth" and his reliance on the "underlying pattern from fairy tales," a pattern that to readers reflects "a truth deeper than historical accuracy" (1990:61). Also see Steven Pinker's comments on the unreliability of autobiographical statements by scientists; "when asked to submit an essay about our lives, we become content providers who edit the events into the satisfying arc of a good plot" (2005:84).

10   Stewart Guthrie (1993) has proposed that an inclination to anthropomorphize is adaptive and part of our mental hard-wiring. Even many of those who would question this strong claim would not challenge the broader assumption implied in it: that anthropomorphizing is a universal human inclination.

11   As alluded to earlier, I am well aware of voices proclaiming that science is culturally and/or ideologically infused on all levels; for these, the dialectic I describe here between "findings of science" and "mythologizing" may appear to be, rather, a dialectic between different levels or sites in the production of culture or ideology under the rubric of science. In asserting a dialectic of "findings" and "mythologizing," I am not proclaiming adherence to the idea that a culture- or ideology-free science is either possible or impossible. My claim is more modest and less final: I am saying that there is a difference between an investigation that is self-critical, rigorous, and circumspect in drawing larger inferences from limited observations, on one hand, and one that seizes on limited observations to expand on moral and political insights and launch grand humanistic and aesthetic visions, on the other.

12   Classical Greek philosophers long ago initiated the dialectic of myth and its would-be heir by offering moments of synthesis amidst the opposition; for an elementary example, one need think no further than Plato's allegory of the cave, which dramatizes the origin of philosophy through rather standard mythological imagery of the passage from dark to light and from subterranean to earthly abode.

At least in the popular realm, the opposition of myth and philosophy has been largely replaced by the opposition of myth and science. In *The*

Fabric of the Cosmos (2004), physicist Brian Greene offers an updated
variant of Plato's allegory of the cave. Recalling his own youthful enthusiasm for science, Greene's elaborate variant begins thusly: "I remember
thinking that if our species dwelled in cavernous outcroppings buried deep
underground and so had yet to discover the earth's surface, brilliant sunlight, an ocean breeze, and the stars that lie beyond ... our appraisal of life
would be thoroughly compromised" (2004:4). This image occurs as part
of the extended rumination on Albert Camus' *The Myth of Sisyphus* with
which Greene opens his discussion of string theory in contemporary scientific cosmology. In his own way Greene embraces the familiar refrain
that science offers a way out of myth, for "unlike Sisyphus, we don't begin
from scratch. Each generation takes over from the previous ... and pushes
up a little further" (2004:22). Greene opens his recent work on multiverse
theory, *The Hidden Reality* (2011:3) with another childhood experience,
without which "my childhood daydreams might have been very different":
that of mirrors infinitely reflecting off one another.

13   Evocations of mythology in Wilson's *Consilience* are many, complex,
and deserving of further study; they are made that much more intriguing
by appearing within a ponderous manifesto portending the impending
encompassment of all knowledge in science.

    In addition to the idea of science as a "new mythos," we encounter
in *Consilience* an attempt to characterize the scientific method through
mythological serpent imagery and the image of Ariadne's thread (1998:
chapter 5) and an attempt to incorporate mythical-archetype theory into
evolutionary psychology and aesthetics (see especially chapter 10 and
pages 300, 314).

    A fascinating link between Hollywood (where celebrity mythologist
Joseph Campbell's influence is broadly acknowledged) and popular science writing is offered in an endnote to *Consilience*; Wilson (1998:300
[note to page 28]) credits his own "placement of the Enlightenment found-
ers in mythic roles of an epic adventure" to the inspiration of Campbell's *The Hero with a Thousand Faces* and to a book on the application
of Campbell's *Hero* to popular culture (specifically, Christopher Vogler's
The Writer's Journey: Mythic Structures for Screenwriters and Storytellers
[1992]).

14   Discussing the question of the "why" of the universe, Hawking concludes
A Brief History of Time with: "If we find the answer to that, it would
be the ultimate triumph of human reason – for then we would know
the mind of God" (1988:175). For three very different commentaries on

Hawking's closing remark, see Davies (1993), Michael (2000:138ff.), and Dawkins (2006:18ff).

15 Guthrie (1993) has brought together a good deal of interesting material relevant to that longer history.

16 Paul Rabinow, drawing on the work of Paul Ricouer, discusses the theme of the detour in *Reflections on Fieldwork in Morocco* (1977:5).

17 The logic of the fable would here seem to dovetail with totemic logic: noting the diversity of racial/ethnic features in human pantheons, Xenophanes imagines an analogous series of zoomorphic pantheons. Perhaps the essence of totemism is dramatization. When totemic societies choose different animals to stand for particular human clans (the Bear representing Clan A; the Eagle, Clan B; and so on) they are in effect analogizing human *intra*-species differences – cultural differences within one species – to *inter*-species differences, seemingly as a way of magnifying or exaggerating the uniqueness of each human clan. Similarly, Xenophanes, by imagining different biological species (each projecting its own image as gods), exaggerates the variability of representations of divinity – precisely in order to dramatize and make visible the unitary process behind such variation.

18 See especially Aristotle's *Nicomachean Ethics*, Book Ten ([1955]).

19 E.B. Tylor, the nineteenth-century anthropologist and mythologist most influential in shaping the myth/science contrast, follows in the footsteps of Fontenelle. Those with an interest in such theories may notice in my book a tacit juxtaposing of the ideas of Tylor with those of the French anthropologist Claude Lévi-Strauss, arguably the most influential theorist of myth in the middle/late twentieth century. My interest in Lévi-Strauss lies not in his theory that myth is structured in "binary oppositions" (which for a time was all the rage in academia), but rather in the sophisticated and less explored commentary Lévi-Strauss gives us on the relationship of myth and science – much of which offers provocative challenges to Tylor's views on the matter.

20 See, for example, the polemical pieces scattered through Dawkins' *A Devil's Chaplain* (2003:135ff, 156ff., 242ff.).

21 In popular science writing, "myth" tends to refer to the vast pool of pre- or non-scientific ideas, throughout the world's history, that have arisen as attempts to explain matters now addressed by science. The term "religion" often has a more immediate reference, i.e., to specific ideas or doctrines that constitute devotional beliefs for at least some members of the reading audience. Some of the strategies I consider under the claim that mythologizing makes science more appealing, may occur as well within

the tense arena of negotiating specific conflicts between science and contemporary religious belief. Nelkin (2004), for example, notes that religious metaphors, concepts, and ultimate explanations are often tapped by biologists to allay ethical and religious concerns about contemporary biological research and to convince the broader audience of its importance.

22   See Doug Russell's "Popularization and the Challenge to Science-Centrism in the 1930s" (1993). On "metrocentrism," see Edward Said (1993:48, passim).

The "anthropo-" root (which in Greek is male-biased) arguably introduces a "phallocentric" slant in the naming of the problem. Along with the imperfect terms "popular science" and "science writing" (see the Introduction), I am reluctantly staying with "anthropocentric" and "anthropomorphic" because they are the terms of the literature and for the moment what we have.

23   Naturalizing as a ploy could theoretically be implemented within either a conception of nature filled with gods or a conception of nature that has dispensed with them. With a focus on the modern world, Barthes' *Mythologies* is more about the latter possibility, and for purposes of the arguments presented in *Mythologies* it is fair to elide naturalizing with a de-deified conception of nature.

24   The classic study of the scapegoat in traditional mythology and religion is René Girard's *The Scapegoat* (1986).

25   Other elements of Barthes' concept of myth as well are historically quite old. For example, the political use of myth by some human constituencies to solidify their own social position forms part of Plato's concept of myth as a "noble lie" told by rulers (*Republic* III.414 [1984:112–13]); and the notion that religious hierarchies used their supernatural doctrines specifically to mollify and control gullible masses was a favourite theme of philosophes in the era of Fontenelle.

CHAPTER TWO

1   The possible exception is heroes from the history of science upon whom the scientific community itself has conferred mythic stature.

2   The references in my analysis are to the 1995 edition. The enlarged edition adds considerable new, often very interesting, material, but this does not appreciably affect the arguments that I consider.

3   See Barrow and Tipler, *The Anthropic Cosmological Principle* (1988). Paul Davies' *Cosmic Jackpot: Why Our Universe is Just Right for Life* (2007) offers a discussion of the several variations on the anthropic principle and an update on debates about these and related issues. Not surprisingly,

both Barrow and Davies are winners of the Templeton Prize for Progress Toward Research or Discoveries about Spiritual Realities (Barrow in 2006, Davies in 1995).

4 Homeric-age Greek poet Hesiod, in his *Theogony* (116–53 [1953:56]), gives *Chaos* as the first element or state of the universe. Primordial conditions that are somehow privative, unformed, or disordered – although sometimes also full of potency – are common in world mythology.

5 In Maori mythology, for example, winds and storms originate with the rebellion of the last-born Tāwhiri, who revolts against his older brothers (see Schrempp 1992:58ff.). See also Watson's (1984:300ff., passim.) discussions of weather mythology and Kerry Emanuel's *Divine Wind: The History and Science of Hurricane* (2005), whose author, according to one reviewer, "truly unites the two cultures by bringing scientific aspects of atmospheric physics together with relevant artworks, poetry, fiction and history" (Pasachoff 2007:15–16).

6 A classic study of the motif of the perverted message as the origin of mortality also comes to us from Frazer (1984). Even though particularly taken to heart by myth, the theme of disproportionate consequences is expressed as well in other varieties of folk wisdom. For example, to explain the principle of "sensitive dependence," James Gleick, in his work *Chaos: The Making of a New Science*, invokes the familiar folk ditty of compounding consequence that tells how a kingdom was lost "all for the want of a nail" (1987:23).

7 See my "Mathematics and Traditional Cosmology: Notes on Four Encounters" (Schrempp 1998b).

8 Bruno Latour speaks to both sides of the polarity that I experienced: the fact that the image was a highly abstract representation, on one hand, and that this representation was offered with the air of bedrock reality, on the other. Comparing them to religious images, Latour says of scientific images: "for most people, they are not even images, but the world itself" (2002:19).

   Although most of the arguments to be considered in the chapters that follow have been put forward through the medium of prose, visual imagery plays a crucial role at some points, most notably in the cosmic graphs of Barrow and Gould (chapters 2 and 3), and the visual spectacle of earthrise that provides the focus of chapter 6. No doubt a sign of our times, visual imagery and persuasion also figure with increasing prominence in contemporary theories of myth.

9 An oft-cited source of this line of size-theorizing is J.B.S. Haldane's essay "On Being the Right Size" (included in his *Possible Worlds and Other Essays* [1932:18–26]). Haldane argues that there is a best size for every

animal and human institution; this is referred to by some as Haldane's principle. Various other examples of why specific fantasies are scientifically possible or impossible can be found in Krauss' *The Physics of Star Trek* (1996), Goswami and Goswami's *The Cosmic Dancers: Exploring the Physics of Science Fiction* (1983), Kaku's *Physics of the Impossible* (2008), and scattered through Achenbach's *Why Things Are & Why Things Aren't* (1996). Reversals are also possible: Highfield's *The Science of Harry Potter* (2002), for example, attempts to show how things and events presented as fantasy might be made possible through emerging technologies.

10   See also Donald Brownlee's and Robert Ward's *Rare Earth* (2000). These scientists use the narrow range of conditions necessary to sustain life to argue against the likelihood of complex life forms in other parts of the universe. For a view challenging the rarity of Earths, see Alan Boss' *The Crowded Universe* (2009).

11   See Richard Feynman (1997:73) on the general theory of relativity according to "cocktail-party philosophers."

12   See especially Barrow's (1995) introduction and conclusion; also 58, 243ff.

13   The set of values that must be maintained – not too small, not too large, but "just right" – is sometimes referred to as the "Goldilocks factor" (see, for example, Davies 2007: passim.).

14   A quirk of Barrow's presentation, however, is that these arguments about non-replicability of structures at different scales follow directly after an enthusiastic discussion of the many fractal or "self-similar" structures in nature. Fractals are "a basic pattern over and over again, each time on a smaller and smaller scale" (1995:59). The relation of these two basic possibilities – structures that *are* vs. those that *are not* replicable at different scales – is left unremarked by Barrow, who uses both to the same end: that of dramatizing the constancy of nature.

15   A larger discussion of strategies of "plausible deniability" in science writing is presented in chapter 6.

16   That the nominally scientific view here ends up as a literalizing of myth, rather than the allegorizing or metaphorizing that more typically results when science meets myth, only adds to the perverseness of the project.

17   Since size is the central concept in the chapter under consideration, it should also be noted that Barrow uses this term equivocally. In the opening of the chapter and at some points along the way, "size" denotes physical dimensionality in a neutral sense, with the implication equally that things can be too large *or* too small to accomplish given tasks. At

other times "size" means *being large enough* for a certain task (as in the everyday usage, "he doesn't have the size to be a football player"; one is less likely to hear, "this doesn't have the size [smallness] to be a bacterium") – as if the term itself is slanted toward the large. Barrow follows his fire arguments with a discussion of the general disadvantages that very large size holds for living things; along with these disadvantages, the equivocal usage of the term "size" itself could facilitate a reader's taking away the impression that we are the optimal size for domesticating fire, even though Barrow's fire arguments technically address only our being large enough for this task.

18 On the history of different senses of "culture" (and of "culture" vs. "cultures"), see Stocking (1968, especially chapters 4 and 9).

19 Stephen Hawking offers an amusing speculation on life in two dimensions (1988:164).

20 The flaws in Barrow's analysis that I lay out in this chapter have to do with what is possible in the universe that we know to exist, and thus do not require delving into multiverse theory, though it is of course tempting to speculate about the further complexities that the latter might add. Toward the latter, Barrow voices considerable skepticism, claiming that "the multiverse scenario was suggested by some cosmologists as a way to avoid the conclusion that the Universe was specially designed for life by a Grand Designer" (2005:53). As if in response, Hawking and Mlodinow, in *The Grand Design*, say of the multitude of universes predicated by "M-theory" that, "Only a very few would allow creatures like us to exist. Thus our presence selects out from this vast array only those universes that are compatible with our existence. Although we are puny and insignificant on the scale of the cosmos, this makes us in a sense the lords of creation" (2010:9). Although this comment opposes Barrow's views in one sense, it converges with them in another: other universes are seen as confirming our unique cosmic place, not as opening other possibilities that might challenge it; other universes increase, rather than decrease, our dominion.

21 The problem can also be phrased through the notion of "emergent properties," a term that refers to capacities that arise – in ways we do not yet entirely understand and cannot entirely predict – out of combinations of elements and that are absent in any of the elements considered in themselves. The classic example is "consciousness" or "mind" emerging from the biochemistry of the brain. If emergent properties are not obvious in constitutive materials, and if cosmic evolution is contingent, what is the procedure by which we determine that all possibilities are exhausted?

22 Another Promethean twist to risk management is adduced by Peter Bernstein who, in his popular work on theory of risk, *Against the Gods*, identifies as Promethean the thinkers who developed techniques of risk management: "like Prometheus, they defied the gods and probed the darkness in search of the light that converted the future from an enemy into an opportunity" (1996:1).

23 I am, of course, aware of the obvious potential for anthropomorphic bias in my own argument, for evolutionary process is not a thief trying to slip an alterative fire-maker past us; and furthermore, conceptualizing the cosmos on the model of a "home" is alleged to be the prime source of error in pre-Copernican thought. What I see as analogous is merely the limiting effect of prosaic thinking in the two cases, home security and cosmic evolution.

24 In arguing that humans aesthetically favour savannah environments, the context suggests that Barrow views this allegedly universal preference not as the only possible outcome of any possible course of mental evolution, but only of the course of evolution that actually did occur on this planet (see 1995:91ff.). But in his discussion of the fire-maker, there is nothing to suggest that the context of Barrow's claim is anything less than all possible courses of cosmic evolution.

25 See Marshall Sahlins' (1976: esp. chapter 2) criticisms of Rappaport and his broader tradition.

26 That he evinces little visible interest in mythico-religious parallels or convergences within his nominally scientific arguments does not, of course, necessarily mean that Barrow is free of such influences. There is no shortage of commentary, much of it speculative, about religious impetuses behind anthropically inclined cosmological theories.

27 The "concrete" notion of culture, considered earlier, and the "epicurean" notion of culture considered here, have deep affinities, including the sense of culture as a single, accumulating entity that improves through time. Culture in the concrete is a history of human achievement, and a cultured person in the epicurean sense is an individual with a taste for the greatest of those achievements in the realm of taste.

When confronted with the fact that the imagery scientists tap for metaphors about specific phenomena often overlaps with the imagery found in mythological portrayals of the same phenomena, defenders of science are quick to assert a distinction between the figural, heuristic intention of scientific metaphor and the innocently literal belief allegedly characteristic of myth. From this point of view, it is interesting to find Barrow so firmly in the literalist camp regarding the relation of culture to "taste." Issues of myth, metaphor, and literalism are be discussed further in chapter 5.

28 The same transition is characteristic of Barrow's accounts of virtually all artistic preferences, including savannah-like landscapes, symmetry, and fractal design. Barrow's analysis of each of these preferences begins by citing some sort of evolutionary survival advantage offered by it at some point in our evolutionary history. But having claimed evolutionary advantage as the origin of such alleged universal artistic preferences, Barrow has little to say regarding whether art itself is to be explained in terms of evolutionary advantage; mostly, he seems to say, such preferences are deployed in order to create objects that "look nice." Compare Barrow on this point to another recent work in evolutionary aesthetics, Denis Dutton's *The Art Instinct* (2009). After making a long case for the evolution of aesthetic impulses through Darwinian sexual selection, Dutton concludes with the claim that we need not feel "doomed" by this legacy: such impulses can be resisted and/or revalued through intellect and contemplation (2009:161–2).

29 Note (in the passage quoted above [57–8]) that despite the studio-shelter habitation, Barrow also assumes that a "division of labor" is entailed in cooking (1995:66).

30 This illustration, of course, takes poetic liberty: due to the impossibility of books at a small size, this bit of quantitative wisdom might have to be expressed through a medium other than a photograph in a book. The majority of images that I came upon in searching for an ant bearing a load show a *line of ants with loads*, thus illustrating not only the load-carrying capacity but also the socio-centric orientation of small beings. Barrow's photograph is of a single ant, although a partially visible second ant (cut off by the left border) might suggest a larger context of collective effort.

31 H. Floris Cohen (1994:86ff.) discusses the theme of quantitative precision in relationship to the growth of science, including Alexandre Koyré's view of a transition from a worldview of "more-or-less" to a "universe of precision." The "quantitative impressionism" to which I call attention in contemporary popular science writing makes liberal use of the devices, symbols, and terminology of precise scientific quantification. But often these are implemented quite laxly, sometimes in cases for which actual measurements do not exist – as if to suggest that the measurements, if they did exist, would confirm the theorist's impressions. Thus, contemporary quantitative impressionism marks a reversion from precision to more-or-less neither in theory nor in worldview, but only in practice.

32 The wonderful image of savoury barbecue – this more immediate and ele-
    mental "taste-test" confirmation of our evolutionary path – has recently
    received an unexpected seconding appraisal, from high authority, via a
    recently published book by Richard Wrangham (*Catching Fire: How
    Cooking Made Us Human* [2009]). The second praise-blurb on the back
    of this new book is by celebrity grill-master Steven Raichlen, author of
    The Barbecue Bible and BBQ USA and host of public television series *Pri-
    mal Grill* and *Barbecue University*, who says: "how did we evolve from
    Australopithecus to *Homo sapiens*, and what makes us human? The
    answer can be found at your barbecue grill and I dare say it will surprise
    you" (Raichlen quoted on Wrangham [2009] back cover). The endorse-
    ment may be more than a gimmick, for Wrangham, like Barrow, sees
    gastronomic taste as holding clues to our evolutionary history. For my
    response to Wrangham's book, see "Catching Wrangham: On the Mythol-
    ogy and the Science of Fire, Cooking and Becoming Human" (Schrempp,
    2011). Some of the arguments about cooking presented by Barrow are
    paralleled in Wrangham, but they take on a highly dissimilar character by
    virtue of an underlying difference in slant: briefly, Wrangham is arguing
    within the metaphysical category of "actuality," while Barrow is arguing
    with the category of "possibility." The two types of persuasion employed
    by Barrow in regard to cooking – the one appealing to quantitative mys-
    tique, the other to an immediate, visceral response – are paralleled by the
    two Copernican revolutions that are discussed in chapter 6.

33 In *The Super-organism*, Bert Hölldobler and E.O. Wilson note that insect
    societies "illustrate, through thousands of examples, how the division of
    labor can be crafted with flexible behaviour programs to achieve an opti-
    mal efficiency of a working group. Their networks of cooperating individ-
    uals have suggested new designs in computers and shed light on how neur-
    ons of the brain might interact in the creation of mind" (2009:xviii).
        The Super-organism would be a rich source of thought-provoking infor-
    mation for one attempting to imagine small fire-domesticators.

34 That which epitomizes also hierarchically encompasses (though of course
    in another, less interesting sense, it is the cosmos that encompasses the
    hearth, since the hearth is *in* the cosmos). On encompassment as hierarchy
    see Dumont (1980). See also geographer Yi-Fu Tuan's *Cosmos & Hearth*
    (1996) and Fustel de Coulanges' classic, *The Ancient City* (1877); these
    works offer fascinating observations on the hearth in human life and the
    construction of social hierarchies, indeed entire social orders, around it.

35 However, since the idea of the cosmos as "full" – as a *plenum* – is wide-
    spread in mythology, the "chain of being" is arguably a philosophical

inflection of an originally mythological idea (see Schrempp 1992;1998b). In some eighteenth-century versions of the chain, pre-ordained essences were said to unfold over time. Lovejoy ends his history with the onset of Darwinian thought, as though actual "transmutation" of species contradicts the principle of the chain.

I am not the only analyst of popular science writing to raise the issue of lingering effects of the chain of being. Davidson's biography of Carl Sagan, for example, cites the chain of being as one of the motivating principles behind the search for extraterrestrial intelligence (1999:30).

36   The notion of a pre-Copernicanism of time is inspired by Stephen Jay Gould (in arguments considered in the following chapter). The cosmic "hodiecentrism" implicit in the view that all possibilities by now have been actualized can be seen as a variation on the enticing theme that Gould challenges, i.e., that the human species forms the telos of cosmic evolution.

37   In introducing his first graph (reproduced on page 45), Barrow comments, "one might have expected the things of the world to be scattered all over the picture in a completely haphazard fashion, showing that Nature explores all possibilities. Nothing could be farther from the truth" (1995:49). Although this statement might appear to directly challenge the idea of the chain of being, actually it does not. The lack of things scattered haphazardly over the picture is not evidence against the proposition that Nature explores all possibilities; it is evidence only against the proposition that Nature explores impossibilities. On the question of whether Nature explores all possibilities – which in Barrow's case would be the question of whether every kind of thing that could exist on the "line of constant density" actually does exist – Barrow says nothing. See also the following note.

38   Since many entities of the cosmos are not represented, the graphs *do not prove either proposition*. The graphs can only suggest, and they suggest one proposition as easily as the other. To really test either interpretation of the graph (i.e., that all things that are possible, are; or that all things that are, are possible) one would have to *fill out the graph with all things that actually are*. Then if one wanted to see whether the proposition of the chain of being is correct, one would check to see whether any gaps appear between these things. If one wanted to see whether Barrow is correct, one would check to see whether any of these things fall within regions of impossibility.

The "contents" of Barrow's chain and those considered by Lovejoy are not identical: Barrow is concerned less with displaying the range of biological species than the range of magnitudes of different kinds of things (some of which are biological species) found in the universe. But ultimately the chain is less about any set of particular contents than about overarching, abstract principles of arrangement and relationship between entities that form a series. Barrow, Lovejoy, and indeed most theorists and commentators on such principles typically provide a meagre sampling of actual entities, just enough to add a little flesh to their proffered overarching principles; unfortunately, this sketchiness facilitates the possibility of alternative readings. In the following chapter, I discuss a diagram (reproduced on page 77) in which Stephen Jay Gould, through a statistical curve, represents the spectrum of living things arranged according to complexity; note that his graph employs only six organisms to represent this spectrum.

In one sense, Barrow's graphs, because they are mostly empty, seem diametrically opposed to the fullness argument of the chain. They portray a thin line of possibility – the "line of constant density" (1995:51) – running through a large empty space. But the "chain" too is a thin, linear image; and on a deeper level Barrow's diagram is not incongruent with the cosmic "fullness" portrayed by the chain. Classical and medieval proponents of the chain never doubted that there were things that could be mentally entertained but could not be; the difference between the Middle Ages and Barrow on this point is that the realm of the ruled-out in the Middle Ages was defined by laws of logic rather than constants of nature: a round-square and an imperfect God are candidates for things that we can name but that cannot actually be (see Lovejoy 1960, chapter 3, on medieval theological antinomies). The chain did not assert that impossible things are, only that *of things that can be*, all are. See also the previous note.

39 See chapter 1 and also Fontenelle, *Origin of Fables* (1972).

40 *Totem and Taboo* (1950) is Freud's classic statement of the origin of culture in the renunciation of incestuous sexuality. Less well known is Freud's own version of the fire myth in *Civilization and Its Discontents*; here Freud offers a scenario of the growth of culture from another type of sexual renunciation, specifically, the homosexually tinged desire to put out fire by urinating on it. "The first person to renounce this desire and spare the fire was able to carry it off with him and subdue it to his own use. By damping down the fire of his own sexual excitation, he had tamed the natural force of fire. This great cultural conquest was thus the reward for his renunciation of instinct. Further, it is as though woman had been appointed guardian of the fire which was held captive on the domestic hearth, because her anatomy made it impossible for her to yield to the

temptation of this desire" (1964b:90n1). Freud (1960) also elaborated this theory through a detailed analysis of the Prometheus myth. Despite the enormous difference between their fire myths, Barrow and Freud are both concerned with the anatomical prerequisites for originating culture. For Barrow, this means a species-anatomy adequate for tending a fire; for Freud, it means a gender-anatomy suitable for urinating on fire and thus for renouncing the impulse to do so – culture amounting to just such renunciation, and thus open in heroic form pre-eminently to the male.

41 The renowned late nineteenth-century/early twentieth-century mytholo-gist James Frazer as well concurs on this point; indeed the epigraph from Frazer that opens this chapter continues, "while myths never explain the facts which they attempt to elucidate, they incidentally throw light on the mental condition of the men who invented or believed them; and, after all, the mind of man is not less worthy of investigation than the phenom-ena of nature, from which, indeed, it cannot be ultimately discriminated" (1930:1).

42 Particularly controversial is a notorious and enigmatic algebraic formula that Lévi-Strauss offers (1967:225) as the universal formula of all myths.

43 The clearest statement of this new theme in Lévi-Strauss is found in the Preface to the Second Edition of his *The Elementary Structures of Kinship* (1969b: see especially xxix ff.).

44 Another way in which Lévi-Strauss portrays the process of mythical think-ing – as a process of *bricolage*, or a handyman's puttering – also suggests a closed quality; the metaphor is of a handyman who accomplishes dif-ferent tasks with a set of tools that, while subject to ingenious variation in application, remain ultimately fixed (1970:16). By contrast, science, for Lévi-Strauss, implies an approach to understanding that can be radically self-transcending in its tool (i.e., conceptual) kit. Thus Lévi-Strauss' notion of "myth" sometimes foregrounds the rather traditional connotation of unchallengeable dogma rooted in social functionality – the fact that myths often justify and serve to maintain structures of order and power within societies.

The "closed" vs. "open" axis is not an entirely satisfactory criterion for distinguishing myth and science, but is adopted for immediate purposes in this instance since "closed" accurately describes at least one strand or ten-dency within mythological thought – a tendency, moreover, that popular science writers, too, seem to often have in mind when they contrast myth to science.

45 In his *Mythologies* (1995), Roland Barthes gives us the quintessential analysis of bourgeois contentment as modern mythology. He argues that

this mythology operates through a naturalizing discourse – presenting a Eurocentric world as though it is given in nature rather than as a consequence of contingent history and human volition (see the fuller discussion of Barthes in chapter 1). It is difficult to imagine a more ambitious naturalizing gesture than Barrow's attempt to root the size of the fire-maker – and smuggled in around its size, its cultural tastes – in the fundamental constants of nature.

CHAPTER THREE

1   It is worth noting that Gould (1997:141), in discussing Darwin, displays a liberal attitude toward contradiction, seeing it as a possible manifestation of intellectual greatness.

2   Gould notes that the left wall may in some cases be an "artifact of an arbitrary human decision" (1997:157) based on the fact that planktonic forams are gathered through sieves that stop only forams above an arbitrary size, allowing the others to go down the drain.

   Gould's focus here on a wall that defines the minimum size of living things has obvious methodological overlap with Barrow's concern with such walls (including the walls that define the possibility of a fire-maker) as discussed in the previous chapter; I present a brief, summary comparison of Barrow and Gould at the conclusion of this chapter.

3   Note that both Gould and Barrow (in the previous chapter), though in different ways, enthusiastically invoke the idea of self-similarity; yet both are imprecise in defining the conditions under which this principle finds application.

4   For a very readable example of an athletic ritual as cosmic microcosm, see Nabokov's (1981) discussion of the tradition of Native American foot races (cf. Schrempp 1992:170ff.).

5   In asserting that "I am not a measure of central tendency ... I am one single human being" Gould in effect is rebelling against the essential process that defines statistics, namely, the stripping of unique identities in favour of overall tendencies. Gould's personal rebellion does not prevent him from developing his ultimate cosmic lesson, about the human place in the cosmos, precisely through a statistically inspired argument. Compare the sentiment to the one offered in computer visionary Jaron Lanier's recent title, *You Are Not a Gadget* (2010). Also compare to Richard Dawkins' more cosmically inflected rebellion in chapter 11 ("Memes: the New Replicators") of *The Selfish Gene*, in which he expands the domain of "replicators" from gene to meme, from nature to culture, only to conclude, famously, "we are built as gene machines and cultured as meme machines,

but we have the power to turn against our creators. We, alone on earth, can rebel against the tyranny of the selfish replicators" (1989:201). As with Gould and his statistical analyses, Dawkins is turning not so much against his creators as his own creation – especially so regarding the concept of the meme. Creators rebelling against, or selectively rejecting, their own creations – the inverse of creatures rebelling against creators – is not an uncommon theme in traditional mythologies. In some instantiations of this theme (notably in Native American traditions of southwestern North America and Central America, including the Mayan *Popol Vuh*), creators casting off of their failed creations may form part of the process of the world – like an inventor's workshop – acquiring its diverse inventory of beings.

6   No doubt human desire (or drivenness in the psychological sense) often enters the picture at some level even for objects whose exquisiteness lies outside of any power we have over, or within, these objects; however, our striving, for example, to view spectacular sunsets or to acquire diamonds does not enter into the very constitution of these objects in a way that is comparable to the way our striving constitutes and shapes, for example, the game of baseball.

    If not all objects we regard as exquisite are statistically skewed, in their character as objects, by human wilful striving, it is also important to keep in mind that such striving is not the only factor that can cause a distribution to be skewed, nor is it the only possible source of statistical drivenness. Following Gould, the overall situation would seem to be that curves can be skewed by either passive or driven processes, and that wilfull striving is one, but not the only, possible source of the drivenness in those processes that are statistically revealed to be driven.

7   The fact that a doctrine of radical contingency arising in science is perceived as supporting a doctrine of radical necessity in the context of religion surely has something to do with the *why me?* mindset of the latter – in other words, with the latter's rejection of the possibility that events either strikingly fortunate or strikingly unfortunate could be random.

8   See the comments on Xenophanes and the fable genre in chapters 1 and 2. A sophisticated treatment of another fable that became entwined in modern debates about hierarchy, Bernard Mandeville's "Fable of the Bees," is found in Dumont (1977: chapter 5).

9   See especially the discussion of Charles Doolittle Walcott in Gould's chapter 4.

10   The great social science work on right and left is by Robert Hertz (1973), a student of Emile Durkheim. Another great work on dualism in ethical reasoning, also influenced by Durkheim, is Henri Bergson's *The Two*

Sources of Morality and Religion (1954). The two sources that Bergson identifies partially resonate with the dualistic (right tail vs. mode) valuations that I have called to attention in Gould; i.e., for Bergson, one source is social solidarity, the other the appeal of a "higher" teaching.

I teach in a department of folklore, a discipline in perpetual quandary over issues of populism and elitism. The term "folklore" frequently connotes something like the wisdom and artistry of everyday life in contrast to the wisdom of cosmopolitan sciences and specialized academic disciplines. Specializing in the wisdom of the mode within an institution that considers itself epistemologically right tail is a source of endless and varied angst among folklorists.

11 Opening a scientific essay by noting the work of art or architecture that had triggered these particular musings forms a sort of Gouldian signature. For example, see Gould (2003b).

12 My comments should not be taken to imply that Gould regards the sufferers whose survival time was medial as in any way less important individuals than those whose length of survival fell in the right tail. Gould's topic is how we should *think about* cancer survival. The reversal in question here has to do with the relative power of the mode and the tail to influence our sizing-up of a situation – and Gould, in the course of telling his personal story, shifts our focus from modal fatalism to right-tail hopefulness.

13 See the discussion in chapter 2 of the same point in regard to Barrow.

14 Gould invokes "mythology" (and related terms such as "myth," "origin myths," and "creation myths") not only in reference to stories about the origin of Earth, the cosmos, and the human species; but also in reference to ideas about recent phenomena, especially baseball (see Gould's "The Creation Myths of Cooperstown" [1991]; cf. Gould 1997:32). In the latter case as much as the former, "mythology" and its derivatives carry a distinct connotation of falsity.

15 See Plato's *Republic*, Book III.

16 This allegory opens Book VII of *The Republic*.

17 On the power of personal experience narratives see Dolby Stahl (1989). Gould's account of storytelling is only one of many accounts of the ways in which everyday human teleology interferes with scientific objectivity. For another recent account, framed in terms of the cosmic "asymmetry" in relation to the human experience of being an "agent," see Huw Price (1996:168 ff.).

It should be kept in mind that while wilful striving is one factor (and the one most relevant to Gould) that can produce a driven curve, not all driven curves necessarily involve this factor.

18   The potential for aesthetic appeal in randomness is tapped by literary forms ranging from high tragedy to self-help (in which "stop blam-
ing yourself!" is a common theme). Gould's and Mlodinow's choices of
title present an interesting reversal: Gould chooses a driven image for his
book title but ultimately delivers a moral message about the necessity
to accept the passive process that assigns our place in nature. Inversely,
Mlodinow chooses a passive image for his book title but ultimately delivers a moral message (one converging with Gould's lead-in examples) advo-
cating, although through a gentle persuasion notably different in tone
from Gould, a patiently driven attitude toward life. Mlodinow concludes:
"What I've learned, above all, is to keep marching forward because the
best news is that since chance does play a role, one important factor in
success *is* under our control: the number of at bats, the number of chances
taken, the number of opportunities seized" (2008:217).

19   The materializing of nature in Lucretius is also accompanied by a mythologizing of the new mode of understanding (that we now call ancient science). Like Gould, who champions Darwin, Lucretius holds up a particular "culture hero" – specifically, the Greek philosopher Epicurus, whom he
praises in epic if not theological terms: "Who has such mastery of words
that he could praise as he deserves the man who produced such treasures
from his breast and bequeathed them to us? No one, I believe, whose body
is of mortal growth. If I am to suit my language to the majesty of his rev-
elations, he was a god – a god indeed ... who first discovered that rule of
life that now is called *philosophy*, who by his art rescued life from such
a stormy sea, so black a night, and steered it into such a calm and sun-lit
haven" (1951:171; cf. Schrempp 1996).

20   Taken at face value, Zeno seems to argue that motion anywhere in the
cosmos is impossible, thus negating the possibility of the race that leads us
into the analysis. One version of his argument is that by the time Achilles
reaches the starting point of the tortoise, the tortoise will have progressed
to a new point; and by the time Achilles reaches that new point, the tortoise will have again progressed ... and so on ad infinitum; thus Achilles
can never overtake the tortoise. The paradoxes of Zeno can be found in
Lee (1967); see also my commentary on Zeno's paradoxes, his allusions
to fables, and athletics as an arena of cosmological thought (Schrempp
1992).

The Zeno problem shows up in a recent work of popular science writing that once again connects the paradox with athletics – John Brenkus'
The Perfection Point (2010:234), which deals with sports science methods
for calculating the absolute limits of performance, including the longest
home run. Curiously (given the topic of Brenkus' book), the paradox, of

which there are many versions, is introduced not through the fabled race of Achilles and the tortoise, but rather through the scenario of a prince trying to reach the king's daughter and being foiled by halfway points – although of course the latter scenario offers its own kind of compelling image of the unattainable. Brenkus might be seen as presenting the scientific counterpart to Gould's rhetoric regarding the perfection point or the possibility of actually touching the wall; here Gould seems to prefer that there be room left for "alchemy" (see above). A brainteaser inherent in Brenkus' analysis is that of trying to imagine a theoretically achievable absolute limit without at the same time imagining the possibility of surpassing it (a sort of Zeno problem in reverse). In the epic discussed just above, Lurcretius at one point invokes something like this teaser to argue for the infinite expanse of the universe, providing another famous, ancient example of a scenario from the sports arena tapped for thinking cosmology. Lucretius asks the reader to imagine a dart thrower hurling his missile at the edge of the universe: does it continue (in which case the universe continues) or does it stop (in which case also the universe continues since there is something lying beyond the border to stop the missile); "with this argument I will pursue you" (1951:55–6). Problems of the infinite and infinitesimal return, in quite different form, in chapter 5.

21 A parallel reversal of the title's meaning occurs in the companion volume, *Wonderful Life* – a title inspired by the Frank Capra film *It's a Wonderful Life*. However, the message of this film seems to be something like *don't give up because, despite bad luck, you will eventually reap what you sow*, while Gould's book is about an opposite type of wonderfulness, that flowing from evolution's wild, accidental, unpredictable course. In the theme of contingency Gould finds a parallel between biological evolution and Capra's film: "This theme is central to the most memorable scene in America's most beloved film – Jimmy Stewart's guardian angel replaying life's tape without him, and demonstrating the awesome power of apparent insignificance in history. Science has dealt poorly with the concept of contingency, but film and literature have always found it fascinating" (1989:14). Gould's analysis of Capra is incomplete and misleading. Yes, small differences beget large ones; but this theme is used by Capra not to dramatize life's contingency but rather the opposite: to show that in the moral sphere, contingency is an illusion brought about by short-term thinking. What is wonderful about life, as portrayed in Capra's film, is precisely that *in the long term it is predictable*: despite the chance calamities, good deeds will eventuate in a broader human good and bad deeds will eventuate in a degraded human condition – though we would have

to be able to know the future to see that the relation between it and the present is non-contingent and non-ironic (hence the film's long flash-for-ward). Such optimism parallels that expressed in the familiar phrase "the judgment of history," as though temporal duration increases the trust-worthiness of moral evaluations.

That Gould is intrigued by Capra is not surprising. Both of these figures are drawn to art and the humanities as well as the sciences. Before his film days Capra studied at the precursor of what is now Caltech, and he later directed a pioneering set of science education films (including the classics *Our Mr Sun* and *Hemo the Magnificent*, which are discussed in chapter 5). While achieving elite social and cultural status, both Gould and Capra were also drawn to popular media and became masters at voicing ebulli-ent populist sentiments.

With *Cosmic Jackpot: Why Our Universe Is Just Right for Life* (2007), Paul Davies tosses yet another gaming metaphor, as book title, into the ring, one that (analogously to Mlodinow, above) reverses Gould's title-image and strategy. That is, since "jackpot" (in contrast to "full house") indicates a kind of gaming process whose outcome is determined by ran-dom, slot-machine luck, Davies' strategy too involves a changeup, but in the opposite direction from Gould's; for Davies' sympathies seem to ultim-ately pull toward a teleological force in cosmic evolution.

CHAPTER FOUR

1   This chapter is dedicated to the memory of David Schneider who, in his courses in the anthropological study of kinship at University of Chicago, constantly challenged the idea of kinship as a system isolated from other dimensions of culture and worldview. David, in some ethereal sense I hope this will remove my "Incomplete" in your course.

2   Jean Dietz Moss discusses this passage from Copernicus in the context of a rhetorical analysis: "His exercise of analogy here, glorifying and per-sonifying the sun, is intended to induce his readers to shift their emotional allegiance from what he considered an erroneous explanation of the cos-mos to a more accurate one. An elegant rhetorical appeal to the emotions such as this in a work devoted to science, although strange to us, cer-tainly would not have surprised his audience" (1993:1). I disagree with this assessment of the source of strangeness in the passage from Coperni-cus. The strangeness does not flow from its status as a work devoted to science – for works by present-day science writers, and especially popular works on cosmology, are replete with appeals to emotion and we do not

find it strange. Whatever strangeness we feel flows rather from the fact that this is a passage *from Copernicus*, to whom we have assigned a definite mythic-hero role: that of leading us out of anthropocentrism.

The solar myth that Copernicus cites is not anthropocentric in the sense of placing Earth in the physical centre of the cosmos, but it is anthropocentric in a way that is equally important, namely, in the assumption that the design of the physical universe confirms human moral values and rules that are encoded in kinship and in religious belief.

3 Cf. Detienne's (1986) discussion of the prejudicial attitudes that have often propelled the idea of "myth."

4 On the solar mythology movement, see Richard Dorson (1965; 1968) and Schrempp (1983).

5 Although Polynesia is geographically very distant from the Mediterranean, numerous variants of the story of separation of Sky and Earth as primal parents of the cosmos occur between these two areas, suggesting that a distant historical connection between the Greek and Maori stories is possible. Numazawa's (1984) survey of this myth gives some indication of its geographical distribution.

6 See especially Jenifer Curnow (1985). There is also a summary and commentary on this account in Schrempp (1992:58–9).

7 Such genealogies raise important issues concerning what we mean by narrative. Genealogy might be thought of, in certain contexts at least, as a limiting case of narrative. Any continuum of human action can, in the recounting, be thickened with lavish detail or thinned to the minimum elements that are capable of conveying a sense of a continuum of world and action, i.e., the names of the main players listed in historical sequence. We might think of genealogy as a highly presupposing narrative, one that, relying on the audience's foreknowledge of details, pares the recital to a minimal image, which can always be fleshed out. Maori genealogies are often prefaced by and punctuated with images of organic development such as the growth and spreading of vines – not unlike the Western inclination to think of genealogies as forming "trees." Numerous examples of Maori cosmogonic genealogies can be found in Elsdon Best (1976).

8 In some Maori accounts it is ambiguous whether the first being of the cosmos is single or dual (see Schrempp 1992:60ff.).

9 For Spencer's response to the Kantians, see especially his appendix "Our Space-Consciousness – A Reply" in his *The Principles of Psychology* (1897:651–69).

10 In both Durkheim/Mauss and Lakoff/Johnson one finds a recognition of *both* a universalist level *and* a culturally relative level of analysis. In both

projects the universal level amounts to a set of categories reminiscent of Aristotle's categories.

But both projects also recognize culturally divergent shapings of fundamental orientations. For example, some societies are organized on the principle of "dual organization" or "moieties" (in which all members of a society belong to one of two macro-clans). Such societies tend to carry this principle of organization into classification in general, so that everything in the cosmos, not just humans, belongs to one of two cosmic clans. Durkheim and Mauss treat the principle of moiety organization as an epistemologically generative social form that is relative to social morphology rather than as a human universal. For their part, Lakoff and Johnson approach "Time is money," or more broadly the view of time as commodity, as a realization of time that not all societies share (1999:163ff).

Kant, Durkheim/Mauss, and Lakoff/Johnson are unified in the fact that they are all convinced of the existence of universals in human knowledge even while insisting that such knowledge in some sense is relative to the nature of the knower.

11 Indeed at some points in *Primitive Classification* it is impossible to tell whether Durkheim and Mauss are talking about the cosmos or the category: they seem to think of the image of society as inscribed simultaneously onto the cosmos – as when a society portrays the spatial organization of the cosmos on the model of the organization of its sub-clans – and onto the form of the category. The data of myth and ritual favoured by Durkheim and Mauss no doubt contributed to oscillation between the category and the cosmos as the object of "this anthropocentrism, which might better be called *sociocentrism*" (1972:86).

12 The contrast here necessitates the use of "anthropomorphize" in the narrow sense, to mean the imposing of human body form. This differs from my general preference for a broad usage (see chapter 1), in which anthropomorphizing could also include the projection, onto non-human objects, of forms that humans have created. The problem in the present instance is that since society is a form created by humans, according to the broad usage sociomorphizing would be part of anthropomorphizing rather than offering a contrast to it (cf. Durkheim and Mauss' comment on "anthropocentrism" and "sociocentrism" in the previous note).

13 We have, specifically, in Freud and in the various divisions of his followers, a great variety of formulations on the general theme that the structure, moral topography, and major "personalities" of the cosmos are projections of experiences based on one or another relation of familial kinship. The posited dynamics can be very intricate; for example, in *Moses*

and Monotheism, Freud implicates the projection of stress points in familial relations not only in accounting for the shape of particular cosmological beliefs, but in accounting for a shift from one belief system to another: "The ambivalence dominating the father-son relationship shows clearly, however, in the final result of the religious innovation. Meant to propitiate the Father Deity, it ends by his being dethroned and set aside. The Mosaic religion had been a Father religion; Christianity became a Son religion. The old God, the Father, took second place; Christ, the son, stood in his stead, just as in those dark times every son had longed to do" (1967:111). Freud's general approach to cosmological projection has given rise to many variants, and it sometimes appears to be complemented by a sort of rebounding projection – of astronomical images into the sphere of psychoanalytical terminology and metaphor. Jacques Lacan is the exemplary case of this tendency, which can be seen, for example, in his discourses on the phallus and the meteor, why planets can't talk, and the meaning of Freud's proclamation of psychoanalysis as a Copernican Revolution (e.g., 1991:3, 13, 16, 224, 234–40). The tendency itself might be seen, from a slightly more distant perspective, as a new variation on an age-old fascination with the possibility of formal parallelism or mutual causal influences between heavenly constellations and human ones (e.g., see Aveni 1994:10ff.). The American anthropological "culture and personality" movement of the mid-twentieth century added to the psychoanalytic movement a heightened sense of cultural relativism by emphasizing the variability of psychodynamic "constellations" comprised in the child-rearing practices of different societies; such variability, it was argued, would be reflected in projective mechanisms, including cosmology (see Kardiner 1945; Spiro 1978).

14   The fact that Lakoff and Johnson here propose a specific "essence" prototype for the clearly bounded category, may have to do with previous work on "fuzzy" categories based on the principle that Wittgenstein dubbed as "family resemblance" (see especially Lakoff 1987). Grouping by "family resemblance" is a possibility that is never considered by Durkheim and Mauss in the pre-Wittgensteinian milieu.

15   As is the case with many works of cognitive science, the neural level is present as a sort of Promised Land toward which researchers are heading. The homage to neurons in Lakoff and Johnson is parallel to that which will be encountered in the following chapter in the context of artificial intelligence research. For although in the latter endeavour computer emulation of intelligent processes offers a way of doing cognitive science without directly studying neurons, such efforts are sometimes accompanied by

the assumption that successful emulations will turn out to have counterparts in neural structures.

16 On the place of Aristotle's categories in the organizational plan of Durkheim's school, see N.J. Allen (2000:32ff. and chapters 5 and 6) and Schrempp (1992:160ff.). Unlike the American anthropology of the same period (in which each student tended to focus on a particular tribe or clan), in Durkheim's program there was a tendency for students to choose one Aristotelian category and pursue it cross-culturally.

17 In their collection of metaphors in *Metaphors We Live By* (Lakoff and Johnson 1980:74ff.), we find very few kinship metaphors, and those we do find are similarly explored from the perspective of their relation to a generic individual. For example, birth metaphors (such as "Our nation was *born out of* a desire for freedom"), which might easily be invoked to emphasize the conceptual depth at which we rely on kinship idioms, and hence a sense of the collective as a model for understanding the world, are assimilated instead to "a gestalt consisting of properties that naturally occur together in our daily experience of performing direct manipulations" (1980:75). The same is broadly true of Mark Turner's exploration of the kinship metaphor in *Death is the Mother of Beauty* (1987).

Everyday, as well as academic, speech is replete with kinship metaphors for categorizing. Staring Lakoff and Johnson in the face is their own repeated invocation of Wittgenstein's "family resemblances" (e.g., 1980:71, 123, 164, 182), yet the phrase – a metaphor that we all know – is not given a significant presence as a datum in their attempts to locate the experiential prototypes of fundamental categories. At one point Lakoff and Johnson themselves create an idiosyncratic metaphor based on kinship: "Classical theories are patriarchs who father many children, most of whom fight incessantly" (1980:53). They announce this as a "novel metaphor," and yet a few pages later they unself-consciously invoke a very standard academic metaphor that is similarly constituted as genealogy ("Examples like these allow us to trace the lineage of our rational argument back ..." [1980:64]). Through such metaphors of the development of our intellectual cosmos, Lakoff and Johnson offer a modern variant of the genealogical cosmos given to us by Hesiod or the Maori. As with Copernicus' kinship cosmos, the reason that kinship figurations (including those that they themselves proffer) slide by these metaphor specialists, escaping their full critical scrutiny, may be that such metaphors are all too familiar (all too familial?).

18 This would also seem to be the message of the book's cover, which bears the image of a single unified mass that appears to be organic (although

it is difficult to tell exactly what the mass is – somebody's backside when posed in the manner of Rodin's sculpture "The Thinker" perhaps).

The differences between Durkheim/Mauss and Lakoff/Johnson as to the generative source of morality offer some interesting parallels to debates in the early twentieth century between Durkheim and anthropologist Bronislaw Malinowski on the perspective of "functionalism," or the idea that social practices should be analyzed in terms of the contribution they make to maintaining a system. Durkheim thought that society should be the telos of the concept of "function," while Malinowski (see especially 1944) thought that the well-being of individuals in society should be the telos.

19 Contemporary assessments of Hertz' work are found in Needham (1973) and McManus (2002:16ff.). The asymmetric dualism of the body is seized upon by Hertz to give voice to a deeper asymmetric dualism that runs throughout Durkheim and all of his students' work, and toward which contemporary cognitive scientists are generally skeptical. Specifically, Durkheim sought to recast the age-old tradition of mind/body dualism sociologically – as the duality of collective vs. individual consciousness. He regarded the duality of individual and society as the most basic duality, the one that underlies the range of specific dualistic formulations found through the world: body vs. mind, matter vs. spirit, sensation vs. intellect, percept vs. concept, the symbolism of left and right. In Durkheim's view, it is in the constitutive nature of society to draw such distinctions, always asymmetrically – one pole predominating over the other just as the social predominates over the individual. Epistemology and moral philosophy derive, according to Durkheim, from the same source: intellect towers over the senses, the concept over the percept, just as the social norm towers over individual desire.

20 Freud's scenario of the original patricidal totemic meal is put forward in *Totem and Taboo* (1950:140ff). See also Girard's (1987) treatment of myths of founding violence, and my (1998b:215ff.) comments on Girard. Lakoff and Johnson develop their claims about embodiment and moral reasoning at the level of a generic ego in its immediate kin relations, and only then do they project these onto larger social-political realms. This part of their approach recalls Freud.

21 There is an interesting parallel between the way Lakoff describes the morality of the Strict Father family and the way that Lakoff and Johnson describe the morality of the body, which is another factor leading me to ask whether, in challenging the Strict Father political model, Lakoff may not also be distancing himself from the body as the source of political morality. Specifically, Lakoff puts strength at the top of Strict Father

family values and empathy below these (1996:379–83; cf. 2008:77–82), a hierarchy that would seem to correlate with the order in which the moral values of the body are listed by Lakoff and Johnson: "health, strength, wealth, purity, control, nurturance, empathy, and so forth" (1999:331; the full text is quoted above). "Nurturance" and "empathy" are last in this list, except for "and so forth." In other words, it appears that body-morality more comfortably fits the Strict Father political model that Lakoff rejects than the Nurturant Parent model that he endorses. This assessment rests on the assumption that the list's ordering implies a hierarchy or, as I have characterized it above, a movement out from a centre; and I do think a reader could reasonably take away that impression. By contrast, a Nurturant Parent family, according to Lakoff, puts empathy and nurturance first (1996:381). I am reading a lot from a little here, but this is unavoidable since Lakoff and Johnson give us little more than the passage quoted above on how to get from kinship morality to the body.

22 The image is ancient; see the Aesopian fable of "the Belly and the Members" at the beginning of the next chapter.

23 See especially his chapter 3 ("The Logic of Simple Freedom").

24 Lakoff (2008:79ff.) suggests a linkage of the binary Aristotelian category to the absolutism of Strict Father (conservative) morality, as discussed above. Seemingly by implication, fuzzy categories belong to Nurturant Parent (liberal) morality.

25 There is a terminological problem in whether one should refer to a Wittgensteinian grouping of family resemblance as a "category." I do follow this usage, a practice that of course leads in the direction of setting up the concept of "category" itself as a fuzzy category.

  Various interrelationships have been suggested between the Aristotelian and Wittgensteinian category; John Taylor (1991:68ff.), for example, presents interesting arguments to the effect that both necessarily have a place in our cognitive life. However, his nomenclature – "folk" vs. "expert" – may be unfortunate, contradicting many academic and nonacademic usages of both terms – including the realm of "high-tech," where, for example, recent computer software has turned to "fuzzy" logic (thus, in Taylor's terms, "folk" logic) in the design of so-called "expert systems."

26 That Wittgenstein chose the term "family resemblance" to stand for just that part of classification that does not conform to the Aristotelian category offers an interesting possible challenge to Durkheim and Mauss, since the latter attempt to derive Aristotelian logic precisely from the kinship of tribal structure. Following Wittgenstein's metaphor, one might argue that Durkheim and Mauss' proposed source offers a bad model of

that for which it is argued to serve as prototype, namely, the Aristotelian category.

However, even if at some levels of focus the difference between Aristotelian categories and Wittgensteinian categories or fuzzy sets is critical, it is not so critical from Durkheim's perspective. In the milieu of social evolutionism, the working assumption was that classification evolved from an original state in which the human mind was able to make no distinctions whatever. Durkheim was attempting to account for the transition from total indistinction to some form of grouping; in this context, the difference between an Aristotelian and Wittgensteinian category is a mere detail.

27 Lévi-Strauss (see 1988:186–7) in the end does not go beyond, nor admit even a possibility of going beyond, the kinship of the clear Aristotelian category, for the "house societies" are seen by him as ultimately not based on kinship, but merely as invoking a rhetoric of kinship.

28 There are long-recognized problems of circularity in Durkheim and Mauss' arguments (e.g., see Needham [1972:xiiff.]). How is the grouping of humans, which will serve as the prototype of the idea of a category, possible without an idea of category in the first place? However, the Durkheimian/Maussian arguments have more plausibility if we envision an ongoing dialectic, rather than a relation of temporal precedence and antecedence. Humans simultaneously deal with different levels and scopes of classifications – from familial relations to universe – and these levels may exercise various kinds of shaping influence on one another.

CHAPTER FIVE

1 This version of the belly and members is recounted by the Roman historian Livy in his *The Early History of Rome* (2.32 [1985:141–2]), who reports its use in a political speech to justify social hierarchy. Livy's account continues, "this fable of the revolt of the body's members Menenius applied to the political situation, pointing out its resemblance to the anger of the populace against the governing class; and so successful was his story that their resentment was mollified." On the antiquity and worldwide distribution of this fable, see Type 293 in Uther (2004:169). Thanks to my colleague Bill Hansen for suggesting this tale and offering a number of insights regarding it. A more immediate historical antecedent to the issues discussed in this chapter is found in the "society of mind" metaphor – based on analogizing the mind as a system of interrelated functions to individuals organized into a society – as it influenced the

"faculty psychology" perspective of the last few centuries (see the previous chapter).

2 Many specializations in several disciplines proclaim an interest in the nature of cognition. For practical purposes (and disavowing the goal of a precise delimitation of the various academic fields involved) I will use two labels: "artificial intelligence" to designate an interest in computer emulation of intelligent processes; and "cognitive theory" when the focus seems to be a more general and broad spread issue in the nature of cognition.

3 This portrayal of the films *Our Mr Sun* and *Hemo the Magnificent* draws on my previous analysis (Schrempp 1998c).

4 These science films also share the buoyant, optimistic tone, culminating in a message of the basic goodness of humanity, characteristic of Capra's other films as well (*It's a Wonderful Life* is discussed in chapter 3).

5 In chapter 1 these two versions, theological and scientific, are characterized in the first and second "moments" respectively.

6 Tylor's guiding epistemological principles are set out in the opening section of Volume One of *Primitive Culture* (1929). On Tylor's epistemology also see Schrempp (1983).

7 Although my implementation is very different from his, my interest in chiasmic relationships between mythic images was ultimately sparked by Lévi-Strauss' development of this theme (e.g., 1969).

8 There are minor exceptions; for example, the phrase "Pandemonium of Homunculi," designating theories that incorporate multiple, competing homunculi, references John Milton's quasi-mythic, pre-modern vision of the capital of Hell in *Paradise Lost*. This is another instance of the power that images of settlement hold in metaphor: Smolin gives us the cosmos as a city full of negotiation and diversity (see preface), while "Pandemonium" gives us the mind as a fractious kingdom.

9 As noted in the discussion of Barrow's fire myth in chapter 2, the domestication of fire and the instigation of the incest taboo are among the archaic events most favoured by nineteenth-century social evolutionists for giving focus to their speculations about the origin and fundamental nature of human society. The origin of the "division of labour" is also such a favoured event, and the subject of many speculative scenarios of the origin of society (recall that Barrow's scenario of the domestication of fire also stressed the division-of-labour required for this achievement).

In describing his methodology, in *Society of Mind*, Minsky says this about his "agents" (i.e., the functional brain parts that Dennett and others refer to as homunculi): "Accordingly, whenever we find that an agent has to do anything complicated, we'll replace it with a subsociety of agents that do simpler things. Because of this, the reader must be prepared to feel a certain sense of loss. When we break things down to their smallest parts, they'll each seem dry as dust at first, as though some essence had been lost" (1986:23).

The passage recalls early sociological debates about the division of labour, especially the angst that swirled around the idea of a transition from a more robust era in which every person was capable of doing every task, to one in which labour became functionally specialized. In the context of increased industrialization, the angst was directed particularly at the dehumanizing effects – the loss of human essence – brought about by specialized rote labour. On the theme of this transition and its attendant moral debates, see especially Durkheim's *The Division of Labor in Society* (1984).

Invoking society to portray mind also offers an interesting reciprocation on a metaphor that figured prominently in the founding of sociology (and which, too, is associated with Emile Durkheim): a body with one mind as a metaphor of society. In the traditional fable of the "belly and members" (epigraph and n1, above), the body also serves as a metaphor of society, although in Livy's version the role of the mind seems to be assumed largely by the belly. See also the discussion of cosmological metaphors of social kinship and of the body in the previous chapter.

10 See especially Maine's *Ancient Law* (1972:100ff.).

11 See the discussion of kinship and body images in mythology and theories of the category, in the previous chapter.

12 Jim Holt (1999) discusses the "double infinity" in historical context. Also see Donna Haraway's discussion of the "equation of Outer Space and Inner Space" in contemporary popular portrayals of science (1991:222ff.).

13 In AI scenarios, even the very top homunculus – whose importance is certainly not less than any other homunculus – is nonetheless a homunculus, suggesting that the brain's physical size – necessarily smaller than that of the body as a whole – has influenced the choice of imagery: the brain part is not any man, but specifically a *little* man.

On the general tendency to spatialize logical/conceptual problems, see the comments on Durkheim and Mauss in the previous chapter; their focus was not interior brain space but exterior tribal space. That Descartes is an object of such fascination – and ridicule – among contemporary cognitive theorists, may have to do in part with the tension between his insistence on a distinct spiritual ("*cogitans*") substance and an equal insistence on theorizing the spatial organic coordinates of our mental abilities.

14 Attneave (1961) offers an oft-cited defence of the project. John Searle thinks it is doomed to failure: "the homunculus fallacy is endemic to computational models of cognition and cannot be removed by the standard recursive decomposition arguments" (1994:226). Sorensen says that according to the theory, "The sequence eventually ends because some homunculi are primitive enough to be explained without subhomunculi – say, merely in terms of feedback mechanisms like a thermostat. Of course, it is one thing to promise this cascade into insentience, another to deliver it. Too often, homuncularists act as if they have discharged their obligation by taking us only a few steps down. We still await a complete homuncular reduction of a psychologically interesting phenomenon" (1992:99).

15 In *The Society of Mind* (1986) Minsky prefers the term "agent" to "homunculus," referring to the latter as "the unproductive and paradoxical idea that a person's behaviour depends upon the behaviour of another person-like entity located deeper inside that person" (1986:329).

   It is interesting to note that in post-structuralist cultural theory, "agent" and "agency" have recently become terms of choice for designating and affirming the operation of distinctly self-conscious, volitional, and personal motives in cultural and historical processes; in other words, these terms have taken on, ironically, precisely those connotations from which Minsky was trying to escape by choosing them over "homunculus."

16 Michel Foucault (1970:17–45) offers, through a comparison of four types of "similitude," an instructive analysis of metaphysical assumptions and hopes that historically have ridden in on the coattails of correspondences discovered or imagined to exist between different levels and/or regions of the natural world. His analysis offers a cautionary message for any inclination to hastily write off cosmic imagery as "mere" analogy.

17 A concern with the question of just how alien we are vis-à-vis the rest of the cosmos arises in different ways and in highly varied works of popular science writing. Compare, for example, the overall effect produced by Barrow's attempts to derive human aesthetic sensibilities from our evolutionary history and the constants of the cosmos (see chapter 2) with Weinberg's vision of our situation as an island of warmth in a hostile, pointless universe (see chapter 6).

18 I discuss Maori/Polynesian counterparts to all of these themes in Magical Arrows (1992:passim.).

19 See the discussions of Margulis and Brisson in the introduction. Torey's book ends with an epigraph that also involves mythic allusion: a passage from Tennyson about "men that strove with Gods" (cited in Torey 2009:209).

20 Comte's delineation of his project can be found in Comte (1975:Part II).

21 Compare to the discussion in chapter 2 of the hierarchy of constraints set
out by Barrow.

22 See my discussions of Gould's (chapter 3) and Sagan's (chapter 6) critiques of such strategic retreats. The upward and inward marches of eight-
eenth- and nineteenth-century socio-cultural evolution were only two of
several, often overlaid, directional schemata. Science progressed upwardly
and inwardly, but also outwardly from Europe to the rest of the world.
For those who followed Hegel, originally it moved from East to West. All
but the most hardened cultural evolutionists also hinted at an opposite,
romantic, downward directionality: with the rise of science, the loss of a
robust, natural poetic innocence embodied in the mythological worldview.
This last directional schema has also been picked up by contemporary
popular science writers (see the discussion of the "paradise lost" theme in
the conclusion).

23 As Cornelia Fales (personal communication) has pointed out to me, we
should also be aware of historical shifts in what is considered mysterious.
An assumption often connected to the idea of the anthropomorphic fallacy is that originally we experienced our minds as familiar and the external cosmos as mysterious; we anthropomorphized the cosmos precisely to
make it more familiar and less mysterious. But when we attempt to displace such anthropomorphizing with a scientific perspective, our minds,
vis-à-vis the rest of the cosmos, emerge as newly mysterious.

24 After presenting a portrait of homunculist logic applied to the tying of
shoes, Fodor says this about the positing of homunculi: "the little man
stands as a representative *pro tem* for psychological faculties which mediate the integration of shoe-tying behaviour by applying information about
how shoes are tied. I know of no correct psychological theory that offers
a specification of these faculties. Assigning psychological functions to little
men makes explicit our inability to provide an account of the mechanisms
that mediate those functions" (1999:47).

25 The clearest example of the two-grid frame is provided by the Freudian-
inspired scheme of Abram Kardiner (1945) constructed around a distinction between "primary institutions" (psychodynamic structures such as the
norms of interaction between parent and child in child rearing), on one
hand, and a realm of "secondary institutions," defined as "projective." Following Kardiner one would expect, for example, that an anxiety-ridden
parent/child relation would be projected cosmically as anxiety-ridden relationships between gods and gods, or humans and gods, in mythology and/
or cosmology. While Kardiner's sense of secondary institutions contains a

variety of likely recipient objects for projections, still the dichotomy of primary or source grid and secondary or projective grid reflects a tendency to think of projection dualistically and unidirectionally.

Interestingly, there is an element of early bidirectionality in E.B. Tylor's theory of "animism," which was enormously important in shaping theories of myth and its relationship to science in the nineteenth century. Tylor thought that the root of animism was the idea of an *interior* substance, ordinarily within but also separable from the body, specifically the "soul." He argued that archaic humans must have been led to this inference by speculating on the difference between a living and a dead body, and by the human shapes appearing in dreams and visions (Tylor 1970:12). We subsequently project the idea of such souls outwardly to account for other aspects of nature; "animism" thus becomes a general theory of the natural world.

26 Tylor's reliance on this formula is discussed by Tambiah (1990:50-1).

27 A provocative work on this issue is Paul Veyne's *Did the Greeks Believe Their Myths?* (1988). Science writers tend to accept what may now be called the standard dichotomy: metaphor is engaged in figuratively and can be heuristically useful; myth is literally held beliefs that have now been disproved by the findings of science. There are only occasional exceptions; for example, Steven Weinberg at one points admits the possibility, regarding the celestial imagery adopted by ancient peoples, that rather than literal belief "they may have just used these theories to suggest convenient markers in the sky, like our own Big Dipper" (2001:177). In the next chapter, I consider an earlier juxtaposing by Weinberg of science and mythology. This earlier instance is more confrontational: myth's unsatisfactory portrayal of the cosmos falls before that of science.

28 Guthrie (1993) argues that our inclination to anthropomorphize offers an evolutionary advantage: when we are unsure of the identity of an unknown object – is it a boulder or a bear? – it is safer to be programmed to bet high. This sort of speculation on evolutionary advantages is common in popular science writing, and often has a "just-so" quality (see also Bird-David's [1999:71] critique of Guthrie's theory). One inclined to do so could come up with evolutionary advantages for introversive anthropomorphizing as well. For example, since the survival of the bodily whole is integrally dependent on the well-being of its parts, an inclination to see the whole in those parts might be evolutionarily advantageous.

29 See the fascinating recent discussion on this topic by Amanda Rees (2001); also see Haraway (1989).

CHAPTER SIX

1   It should be noted that there are occasional popular science writers who, in parallel with some historians of science, discredit the idea of the Copernican Revolution in its heroic form (i.e., as a sudden, one-time, sweeping transformation credited to one thinker). For example, Lee Smolin in the *Life of the Cosmos* and Timothy Ferris in *The Whole She-Bang* both throw cold water on the heroic version, the idea that, in Ferris' words, "the sun-centred Copernican universe brought simplicity and light to cosmology in a single stroke" (1997:24) – though both authors also note the cultural influence of the idea of what Smolin calls the "legend of Copernicus" (1997:3).

2   A more recent popular account is Simon Singh's *Big Bang: The Origin of the Universe* (2004). Noting that Weinberg's book "is still one of the best popular accounts of the Big Bang" (2004:515), Singh begins his own account with a similar myth/science contrast. He summarizes four cosmogonic myths (one from China, one from Iceland, and two from West Africa) telling of creation by "giants." Such myths served as the "final word," until an "outbreak of tolerance among the intelligentsia" in the sixth century BC, at which point philosophers could challenge "accepted mythological explanations" (2004:6). The opening epigraph to Singh's first chapter ("In the Beginning") is attributed to Karl Popper: "Science must begin with myths, and with the criticism of myths" (2004:1).

3   See also the discussion of this principle in chapter 2.

4   Regarding another type of influence of Copernicus on Gould, in this case the Copernicus given to us by Thomas Kuhn, see a *New Yorker* piece by H. Allen Orr, professor of biology at University of Rochester. Specifically, regarding Gould's most notorious theory (introduced with Niles Eldredge), that evolution is characterized not by continuous gradual change but by "punctuated equilibrium," Orr, in an "admittedly speculative" comment, says "Kuhn deeply affected Gould's science. One might even argue that punctuated equilibrium is little more than Kuhn's view of the history of ideas transferred wholesale to the history of life, an idea that is reinforced by the fact that Gould and Eldredge began their 1972 paper with talk of paradigms and Kuhn. Even the title of Gould's magnum opus seems a riff on 'The Structure of Scientific Revolutions'" (Orr 2002:138). Orr's comment is also of direct relevance to the issue of anthropomorphism, since it implies that Gould's scientific theory of evolution originated anthropomorphically, i.e., that Gould derived his model of biological evolution from a model of how we humans interact with and change our ideas. But

there is no need for surprise in the possible anthropomorphic origin of Gould's most important theory, for this would merely be a more striking instance of the process implied in Gould's practice, considered earlier (in chapter 3), of opening his columns by informing his readers of the particular human artifact – often drawn from art and architecture – that has inspired his latest scientific insight about the biological world.

5   As discussed in the previous chapter, Auguste Comte's schema of the orderly evolution of "positive science" offered the most influential formulation of this idea: the first disciplines to become scientific (astronomy and math) dealt with the order of the physical cosmos as a whole, and the next disciplines to become scientific were the biological sciences. The movement will culminate with the social sciences.

6   Olrik's law would seem to be a law of Western culture, not a law of nature (on these laws see Olrik [1965]). Native American traditions usually operate according to a law of four, evidenced, among other things, by European folktales that have been adopted by Native Americans, invariably with the number of obstacles transformed from three to four (for example, see Cushing 1965). See also Alan Dundes' classic essay, "The Number Three in American Culture" (1968).

    The order of Gould's four bullets is also problematic, for the great shock wave from scientific challenges to Biblical chronology actually preceded Darwin, with the geological researches of Sir Charles Lyell as exemplar, so that Gould's fourth bullet is more properly the second.

7   Wolfram's positioning of his work specifically at the pinnacle of the triad comes through even more clearly in a conversation reported by Charles Piller in a front-page *Los Angeles Times* story: "'There just isn't a mechanism within the current structure of science to present things as big as what I'm trying to do,' he [Wolfram] said. Actually, 'big' doesn't begin to capture it, Wolfram says. He describes his theory as 'one of the more important single discoveries in the whole history of theoretical science,' akin to those of Copernicus, who overturned centuries of orthodoxy by proving in 1530 that the Earth was not the centre of the universe, and Charles Darwin, whose 1859 theory of natural selection shattered religious dogma about creation" (Piller 2002:A12).

8   There is another important tradition involved here, one quite beyond the scope of this book, i.e., that of social sciences and humanities marvelling at the progress of the physical sciences and trying to emulate them in order to achieve similar progress. For potent twentieth-century examples of this vision, see Vladimir Propp's *The Morphology of the Folktale* (1968) and Northrop Frye's *Anatomy of Criticism* (1957).

9   See the fascinating material that I. Bernard Cohen has brought together
    in chapter 15 ("Kant's Alleged Copernican Revolution") in *Revolution
    in Science* (1985), and also Blumenberg (1987:595ff.). Cohen calls atten-
    tion to the fact that Kant himself does not use the term "Copernican
    Revolution."

10 The memorable episode is discussed in Simon LeVay's "Space Science: Off
    Target," in his *When Science Goes Wrong* (2008:199–220).

11 See Mailer's account of the liftoff in his *Of a Fire on the Moon* (1970:100).

12 William Shea (2000) gives us an interesting study of the use of earth/moon
    analogies – thinking of Earth as though it were the moon or vice versa –
    in the long history of Western cosmology; from this perspective Sagan's
    project emerges as a new variation on an old analytical/argumentative
    strategy.

13 Existentialist sensibility – philosophical, literary, and popular cultural –
    also demonstrates a particular fondness for the image of wandering or
    journeying, being "on the road." Sagan's Introduction to *Pale Blue Dot* is
    called "Wanderers: An Introduction," and the opening lines are, "We were
    wanderers from the beginning. We knew every stand of tree for a hun-
    dred miles. When the fruits or nuts were ripe, we were there. We followed
    the herds in their annual migrations. We rejoiced in fresh meat" (1994:xi).
    This is quite a roundup of anthropological images (see also the meat
    theme as it figures in Barrow in chapter 2). Sagan's musings continue,
    carrying us from nomadic migrations to seafaring explorers. As if with
    his distant views of the Earth-dot in mind, Sagan notes, "we invest far-off
    places with a certain romance" (1994:xii).

14 Also consider the popularity of such films as *The Right Stuff* and *Apollo
    13*, both set in the heyday of the space program.

15 Presenting us with a stunning picture of the whole earth photographed
    by the final manned moon mission, Sagan says: "While almost everyone
    is taught that the Earth is a sphere with all of us somehow glued to it by
    gravity, the reality of our circumstance did not really begin to sink in until
    the famous frame-filling *Apollo* photograph of the whole Earth – the one
    taken by the *Apollo* 17 astronauts on the last journey of humans to the
    Moon. It has become a kind of icon of our age" (1994:4–5). That Sagan
    calls attention to Apollo 17, the final moon mission, which was followed
    by a long hiatus in such exploration, may in part be an expression of his
    wish that the hiatus end. But even though, for Sagan, the whole-earth
    photo trumps earthrise as the icon of our age, earthrise photos still get tre-
    mendous exposure in *Pale Blue Dot*. Indeed this work contains four differ-

ent full-colour earthrise photos (1994:45, 210, 211, 218), the first of these comprising a sequence of four shots that show Earth at different heights in its "ascent." None of these, however, is the Apollo 8 earthrise that has been culturally canonized as iconic; rather, they are identified as from Apollo 11, 14, and 15.

Denis Cosgrove (1994) presents a rich and fascinating analysis of Apollo photographs of Earth, focusing on the two mentioned above, earthrise from Apollo 8 and the image of the full Earth from Apollo 17. Cosgrove's focus is traditions of geographic, cartographic, and artistic representations and their relation to political and cultural history; other than as suggested in a poetic essay by Archibald MacLeish on the Apollo 8 photos, Cosgrove does not specifically deal with the opportunity for a Copernican lesson from earthrise.

16   The journey also saw some secularized Christmas festivities, including the reading of a parody of "'Twas the Night before Christmas" in mission control and the message "Please be informed there is a Santa Claus" from the astronauts (Brooks, Grimwood, and Swenson 1979:281).

17   For other variants see, for example, Flowers 1998:121; Cernan and Davis 1999:181; Lindsay 2001:189; Lindbergh 1974:xii; and MacLeish 1968.

18   The question of whether this vision was as spontaneous as it is said to be, or whether it may have instead gelled over the years is an intriguing question, one that could be solved only by a micro-level tracing of its manifestations.

19   The photograph of Earth in the sunbeam also appears at another point in *Pale Blue Dot* – conspicuously on the reverse side of the book's title page, with the caption, "The Earth: a pale blue dot in a sunbeam." The format of the photo is elongated, and the beam is now oriented horizontally rather than vertically. In this format the photograph is particularly inspirational, resembling the beam from a lantern or lighthouse, both very common motifs in popular religious art.

A litany similar to Sagan's is offered by Olaf Stapledon; Martin Rees places it as opening epigraph for *Our Cosmic Habitat*. Of the generations of men that have inhabited the "little round grain" of Earth, Stapledon says: "and all their history, with its folk-wanderings, its empires, its philosophies, its proud sciences, its social revolutions, its increasing hunger for community, was but a flicker in one day of the lives of the stars" (as quoted in Rees 2001).

For a roundup of other examples of poetic musings in similar vein see Cosgrove (1994:283ff.).

20   Cosgrove says of the Apollo photos, "The apparent objectivity of the photography and the positioning of the camera so far outside the bounds of Earth seemingly constitute an unchallengeable vantage point" (1994:288).

21   In comments that follow the one quoted above, Lessl characterizes Sagan's motives at two different levels, one psychological and universal ("Sagan is responding to an enduring human need" [1985:184]), the other historical and particular (e.g., "From the millennialism of the old theistic vision *Cosmos* inherits its preoccupation" [1985:184]). Both the psychological and the historical are obviously relevant; the question is how to weigh and interconnect them.

       This issue – the relative weight to give "history" or "tradition" as opposed to universal psychology in accounting for myth – itself has an interesting history in theory of myth, some arguing that mythologies must be understood as products of slow historical accretions, others arguing that the mind is capable of creating myths almost *ex nihilo* – spontaneously and afresh as needed.

22   Another point – or rather, a noteworthy absence – confirms, in a way that is both more intriguing and roundabout, how smitten Blumenberg is by the visual image of the Earth oasis. For Blumenberg fails to mention the lesson of perspectival relativity, noted by Sagan and Stoppard, that is available in the image of earthrise. Blumenberg is among those thinkers who resist the idea of a Copernican Revolution as a sudden, sweeping change; and one of his cherished points of resistance is the claim that Copernicus' insights rest upon a prior intellectual shift, from the notion of Earth as possessing a unique character (vis-à-vis other heavenly bodies) to the acceptance of Earth as part of a homogeneous cosmos. That Blumenberg of all thinkers should fail to mention the possibility of the potential lesson in perspectival relativity offered by the moon missions is noteworthy, for willingness to see Earth's orbit of the sun *as analogous to* the moon's orbit of Earth would dramatize just this presumption of homogeneity.

23   A case in point is Kabbalah, which Primack and Abrams see, as does Torey (discussed earlier), as resonating with ideas from modern cosmological science (2006:197ff.). Primack and Abrams strive to be culturally pluralist by tapping the mythologies of diverse cultural traditions; but consider again the other route, the one taken by Torey who, in effect, pronounces Kabbalah the one true myth (see chapter 5, pages 178–80).

24   That the special qualities of Earth now stand at seven provides a fortuitous resonance with the broader popular sphere, where the ground is already prepared by the quasi-secularized sacred numerology that is a

mainstay of self-help literature – another realm in which readers are seeking "re-centring." Most notable is Stephen Covey's *The 7 Habits of Highly Effective People* (2004), which has inspired an industry of imitative variations that precisely parallel those inspired by the other prominent secularized sacred number, three, in the many variations on the Copernican trinity discussed above. On the other hand, there is an irony: the longer the litany of ways in which Earth turns out to be uniquely habitable, the more the remainder of the cosmos becomes, as alluded to above, alien – seemingly decreasing the possibilities for emotional bonding.

25 Schrempp, "Science Fundamentalism: Problems with Remythologizing the Cosmos," ms. in progress (b). Among the issues to be considered are similarities with and differences from the philosophical tradition of allegorizing of myth, and the tradition of de- and re-mythologizing in modern theology (e.g., see Vanhoozer 2010); and also the pervasive cultural/intellectual influence in our era of mythologist Joseph Campbell, whose presence in Primack and Abrams is even more in evidence than it is in E.O. Wilson.

26 Perhaps the most systematic account of the propensity of cosmological arguments to turn against themselves is Kant's doctrine of the "antinomy of pure reason" (elaborated in the second half of the *Critique of Pure Reason*).

27 For subsequent discussions of Weinberg's comments, see Horgan (1997:73, 244) and Davies (2007:16).

28 See the fuller passage on the implications of multiverse theory from Hawking and Mlodinow in chapter 2, n20.

CONCLUSION

1 The science education film *Our Mr Sun* is considered in chapter 5.

2 Dennett's statement and its context are considered in chapter 5. The statement has an interesting resonance with a statement made by Gould (also about Darwin), which I consider in chapter 3: "if progress is so damned obvious, how shall this elusive notion be defined when ants wreck our picnics" (Gould 1997:145).

3 Sagan 1994:46–7. Though he does not tie it to the "paradise lost" theme (or at least to any strong version of it), Steven Pinker, in *How the Mind Works*, offers another variant of the "then along came science" trope: "and then along came computers: fairy-free, fully exorcised hunks of metal that could not be explained without the full lexicon of mentalistic taboo words" (1997:78). Pinker calls attention to our inclination to

describe the workings of computers with mentalistic terms (such as thinking, knowing, communicating) to launch his defence of the "computational" theory of mind, one part of which is the "homunculist" research
project (see Pinker 1997:79ff.) that I consider in chapter 5.

4   In *Science as Salvation: A Modern Myth and its Meaning*, Mary Midgley
deals with the soteriological theme in science and popular science writing,
opening her discussion with the comment that "the idea that we can reach
salvation through science is ancient and powerful" (1992:1). Thomas
Lessl deals with various themes connected with religious salvation in Carl
Sagan's work, including the themes of original sin and the fall of man
(1985:179–80). What Lessl sees as analogous to the Biblical fall, however,
is Sagan's emphasis on the existence of more evolutionarily primitive parts
in the human brain. Neither of these scholars emphasizes the theme considered here: the idea of an original intellectual state of mixed moral character – blissful security resting on ignorance – a state lost with the dawn
of science and posing a void to be filled.

The study of comparative mythology suggests that the Judeo-Christian
felix culpa is only one among many instances in which a cosmically founding act is steeped in moral ambiguity. For other examples see my discussion of the cosmogonic event of separation of Sky and Earth in Maori cosmogony (Schrempp 1992:60ff.) and the discussion of the myth of Oedipus
by Lévi-Strauss (1967:212ff.), who interprets this myth as posing an
unhappy choice between non-existence and defective existence.

In his *Scientific Mythologies: How Science and Science Fiction Forge
New Religious Beliefs* (2008), James Herrick too deals with the theme of
salvation. A modern Christian apologist, Herrick confronts the proliferation of "redemptive tales" that speculative forms of popular science writing and especially science fiction now offer to the modern audience as
substitutes for the authentic gospel. "This is the Christian church's challenge today – to reclaim its story and tell it in such a way that it stands
out among all the others as authentic, as the Great Story that other stories
have often sought to imitate" (2008:252). There are many problems with
this formulation, most notably that the task of finding the story that is not
(in whole or part) an imitation of another story is all but futile.

5   Benedict's fullest discussion of the theme of "compensatory" functions of
myth occurs in her Introduction to *Zuni Mythology* (1934).

6   Dawkins' discussion of "Consolation" in *The God Delusion* (2006:352ff.)
offers an especially candid treatment of the trade-in theme.

7   But the legend has also begotten a counter-legend: "One of his
[Hawking's] closest friends, David Schramm, has known Hawking for

over twenty years and has little patience with those who try to create an image of Stephen as in any way emotionally different from others. He has never pulled any punches when it comes to his friend's personal life. He once introduced Hawking at a talk he gave in Chicago, by saying, 'as evidenced by the fact that his youngest son Timothy is less than half the age of the disease, clearly not all of Stephen is paralyzed!' Apparently half the audience were shocked speechless, but Hawking loved it" (White and Gribbin 1992:288). As a graduate student, I was a member of the audience at the talk in question and can thus testify to the approximate accuracy of this account.

Discussing journalistic treatments of science, Nelkin (1995:17) points out that "instances of prestigious scientists behaving like ordinary mortals are noted with an air of surprise." Compare this observation to one made by Roland Barthes (in *Mythologies*) concerning journalists' fondness for injecting small elements of everydayness (e.g., reference to an open-necked shirt or print dress, clothes that "can also be seen on the bodies of mere mortals" [1995:32]) into their portrayals of European royalty caught in casual circumstances. Barthes labels this strategy "inoculation" (1995:150): the injection of a small dose of ordinariness in order to protect against serious contamination.

8   Stunning photographs of objects at scales modified by telescopy and microscopy are frequently offered as enticements to the world of science; some imposing works (e.g., *Heaven & Earth* [Malin and Roucoux 2002]), are given over almost entirely to such photographs.

9   Even Richard Dawkins, who built a reputation on tough-minded debunkings of "cosmic sentimentality," comes to argue, in *Unweaving the Rainbow*, for the necessity of "science inspired by a poetic sense of wonder" (1998:ix, xii).

I opened this book by pointing out that Lee Smolin, in discovering the perfect metaphor for his modern cosmological science, also happens to hit upon an ancient mythological conceit: the idea that the human settlement reflects the cosmos. Analogously, Dawkins too, in the waxing-poetic of his maturity, frames his title and argument through an image that is widespread in mythological cosmology, that of weaving (see also Robert Jastrow's use of this image in *The Enchanted Loom: Mind in the Universe* [1981]). Textile metaphors (e.g., "fabric," "wrinkles") are also common in scientific talk of space-time.

10   The *Rules of Sociological Method* (1982) is Durkheim's strained attempt to show that society, in the form of the social "effervescence" generated in collective life, is scientifically measurable.

11 See also mythologist Bruce Lincoln's (1996) reflections on the mythic and ritualistic characteristics of such American immigrant-ancestor stories.

12 Although popular science writers, as noted earlier, are motivated more to save something like the spirit of mythology than to save particular myths, Luc Brisson's study of philosophers saving myths is suggestive. Asking why philosophers of Antiquity and the Renaissance, despite their legacy of skepticism toward myth, were intent to also keep alive the classical mythological heritage, Brisson says: "The answer perhaps lies in that neither reason nor a new faith was better able than those myths to express something very special and irreducible deep in the human heart, as one generation after another ceaselessly and imperceptibly transformed the cultural legacy of ancient Greece and thus ensured its survival" (2004:3;cf. Nagy 2002).

References

Abrams, Nancy, and Joel Primack. 2011. *The New Universe and the Human Future*. New Haven: Yale University Press.

Achenbach, Joel. 1996. *Why Things Are & Why Things Aren't*. New York: Ballantine Books.

Allen, N.J. 2000. *Categories and Classifications: Maussian Reflections on the Social*. New York: Berghahn Books.

Allenby, Richard. 1969. Introduction. In *Analysis of Apollo 8 Photography and Visual Observations*, vii–viii. Washington, DC: Office of Technology Utilization, National Aeronautics and Space Administration. (NASA SP-201)

Aristotle. 1938. *The Categories*. London: William Heinemann Ltd. Translated by Harold P. Cooke.

– 1955. *Ethics*. Harmondsworth: Penguin.

– 1973. *Poetics*. Ann Arbor: University of Michigan Press.

Attneave, Fred. 1961. "In Defense of Homunculi." In *Sensory Communication*, ed. Walter A. Rosenblith, 777–82. Cambridge: MIT Press and New York: John Wiley & Sons, Inc.

Aveni, Anthony. 1994. *Conversing With the Planets*. New York: Kodansha.

Bachelard, Gaston. 1968. *The Psychoanalysis of Fire*. Boston: Beacon Press.

Baddeley, Alan. 1995. "Working Memory." In *The Cognitive Neurosciences*, ed. Michael Gazzaniga, 755–64. Cambridge: MIT Press.

Banner, H. 1957. Mitos dos indios Kayapo. *Revista de Antropologia* 5:1.

Barrow, John D. 1995. *The Artful Universe*. Oxford: Oxford University Press.

– 1998. *Impossibility: The Limits of Science and the Science of Limits*. Oxford: Oxford University Press.

– 2005. *The Artful Universe Expanded*. Oxford: Oxford University Press.

Barrow, John D., and Frank J. Tipler. 1988. *The Anthropic Cosmological Principle*. Oxford: Oxford University Press.

Barthes, Roland. 1995. *Mythologies*. New York: Hill and Wang.

Bauman, Richard, and Charles Briggs. 2003. *Voices of Modernity: Language Ideologies and the Politics of Inequality*. Cambridge: Cambridge University Press.

Beckman, Petr. 1993. *A History of Pi*. New York: Barnes & Noble.

Benedict, Ruth. 1934. *Zuni Mythology*. 2 vols. New York: Columbia University Contributions to Anthropology, No. 21.

Bergson, Henri. 1911. *Creative Evolution*. New York: Henry Holt and Company.

– 1954. *The Two Sources of Morality and Religion*. Garden City: Doubleday.

Bernstein, Peter. 1996. *Against the Gods: The Remarkable Story of Risk*. New York: John Wiley & Sons, Inc.

Best, Elsdon. 1924. *The Maori*, vol. 1. Wellington: Harry H. Tombs.

– 1925. *Tuhoe*. Wellington: Reed.

– 1976. *Maori Religion and Mythology*, part 1. Wellington: Government Printer.

Bird-David, Nurit. 1999. "Animism Revisited." *Current Anthropology* 40:S67-S91.

Blumenberg, Hans. 1987. *The Genesis of the Copernican World*. Cambridge: MIT Press.

Boslough, John. 1989. *Stephen Hawking's Universe*. New York: Avon.

Boss, Alan. 2009. *The Crowded Universe*. New York: Basic Books.

Brenkus, John. 2010. *The Perfection Point*. New York: HarperCollins.

Brisson, Luc. 2004. *How Philosophers Saved Myths*. Chicago: University of Chicago Press.

Brockman, John. 1996. *The Third Culture*. New York: Simon & Schuster.

Brockman, John, ed. 2005. *Curious Minds*. New York: Vintage Books.

Brooks, Courtney, James Grimwood, and Loyd Swenson. 1979. *Chariots for Apollo: A History of Manned Lunar Spacecraft*. Washington, DC: National Aeronautics and Space Administration.

Brownlee, Donald, and Robert Ward. 2000. *Rare Earth*. New York: Copernicus.

Butler, Judith. 2000. *Antigone's Claim*. New York: Columbia University Press.

Cernan, Eugene, and Don Davis. 1999. *The Last Man on the Moon*. New York: St Martin's Press.

Chaikin, Andrew. 1998. *A Man on the Moon: The Voyages of the Apollo Astronauts*. New York: Penguin Books.

Cohen, H. Floris. 1994. *The Scientific Revolution: A Historiographical Inquiry*. Chicago: University of Chicago Press.

Cohen, I. Bernard. 1985. *Revolution in Science*. Cambridge: Harvard University Press.

Cole, K.C. 1999. "The Emotional Pull of Science." *Los Angeles Times*, June 24, B2.

Collins, Francis. 2006. *The Language of God: A Scientist Presents Evidence for Belief*. New York: Free Press.

Comte, Auguste. 1975. *Auguste Comte and Positivism: The Essential Writings*. New York: Harper & Row.

Copernicus. 1978. *On the Revolutions*. Baltimore: Johns Hopkins University Press.

Cosgrove, Denis. 1994. "Contested Global Visions: One-World, Whole-Earth, and the Apollo Space Photographs." *Annals of the Association of American Geographers* 84 (no.2):270–94.

Covey, Stephen. 2004. *The 7 Habits of Highly Effective People*. New York: Free Press.

Cox, George. 1883. *An Introduction to the Science of Comparative Mythology and Folklore*. London: Kegan Paul, Trench & Co.

Curnow, Jenifer. 1985. *Wiremu Maihi Te Rangikaheke: His Life and Work. Journal of the Polynesian Society* 94:97–147.

Cushing, Frank Hamilton. 1965. "The Cock and the Mouse." In Alan Dundes, *The Study of Folklore*, 269–76. Englewood Cliffs: Prentice-Hall.

Davidson, Keay. 1999. *Carl Sagan: A Life*. New York: John Wiley & Sons, Inc.

Davies, Paul. 1993. *The Mind of God: The Scientifc Basis for a Rational World*. New York: Simon & Schuster.

– 2007. *Cosmic Jackpot*. Boston: Houghton Mifflin.

Dawkins, Richard. 1989. *The Selfish Gene*. Oxford: Oxford University Press.

– 1998. *Unweaving the Rainbow*. Boston: Houghton Mifflin.

– 2003. *A Devil's Chaplain*. Boston: Houghton Mifflin.

– 2005. *The Ancestor's Tale*. Boston: Houghton Mifflin.

– 2006. *The God Delusion*. Boston: Houghton Mifflin Company.

– 2009. *The Greatest Show on Earth: The Evidence for Evolution*. London: Bantam.

de Coulanges, Fustel. 1877. *The Ancient City*. Boston: Lee and Shepard.

Dennett, Daniel C. 1978. *Brainstorms*. Montgomery: Bradford Books.
– 1991. *Consciousness Explained*. Boston: Little, Brown, and Company.
– 1996. *Darwin's Dangerous Idea*. New York: Simon & Schuster.
– 2005. *Sweet Dreams: Philosophical Obstacles to a Science of Consciousness*. Cambridge: MIT Press.
Detienne, Marcel. 1986. *The Creation of Mythology*. Chicago: University of Chicago Press.
Dolby Stahl, Sandra. 1989. *Literary Folkloristics and the Personal Narrative*. Bloomington: Indiana University Press.
Dorson, Richard. 1965. "The Eclipse of Solar Mythology." In *Myth: A Symposium*, ed. Thomas Sebeok, 25–63. Bloomington: Indiana University Press.
– 1968. *The British Folklorists*. Chicago: University of Chicago Press.
Drexler, K. Eric. 1986. *Engines of Creation: The Coming Era of Nanotechnology*. Garden City: Doubleday.
Drummond, Lee. 1996. *American Dreamtime: A Cultural Analysis of Popular Movies, and Their Implications for a Science of Humanity*. Lanham: Littlefield Adams Books.
Dumont, Louis. 1977. *From Mandeville to Marx*. Chicago: University of Chicago Press.
– 1980. *Homo Hierarchicus*. Chicago: University of Chicago Press.
Dundes, Alan. 1969. "The Number Three in American Culture." In *Every Man His Way: Readings in Cultural Anthropology*, 401–23. Englewood Cliffs: Prentice-Hall.
Durkheim, Emile. 1965. *The Elementary Forms of the Religious Life*. New York: Free Press.
– 1982. *The Rules of Sociological Method*. New York: The Free Press.
– 1984. *The Division of Labor in Society*. New York: The Free Press.
Durkheim, Emile, and Marcel Mauss. 1972. *Primitive Classification*. Chicago: University of Chicago Press.
Dutton, Denis. 2009. *The Art Instinct*. New York: Bloomsbury Press.
Eger, Martin. 1993. "Hermeneutics and the New Epic of Science." In *The Literature of Science: Perspectives on Popular Scientific Writing*, ed. Murdo William McRae, 186–209. Athens: University of Georgia Press.
Emanuel, Kerry. 2005. *Divine Wind: The History and Science of Hurricanes*. Oxford: Oxford University Press.
Ferris, Timothy. 1997. *The Whole Shebang*. New York: Simon & Schuster.
Ferris, Timothy, ed. 2001. *The Best American Science Writing 2001*. New York: HarperCollins.
Feynman, Richard. 1997. *Six Not-So-Easy Pieces*. Reading: Addison-Wesley Publishing Company, Inc.

Firth, Raymond. 1963. "Bilateral Descent Groups: An Operational Viewpoint." In *Studies in Kinship and Marriage*, ed. I. Schapera, 22–37. London: Occasional Papers of the Royal Anthropological Institute.

Flowers, Charles. 1998. *A Science Odyssey*. New York: William Morrow and Company.

Fodor, Jerry. 1999. An excerpt from "The Appeal to Tacit Knowledge in Psychological Explanation." In *Mind and Cognition: An Anthology*, ed. William Lycan, 46–9. Oxford: Blackwell.

Fontenelle, Bernard. 1972. "Of the Origin of Fables." In *The Rise of Modern Mythology*, eds. Burton Feldman and Robert Richardson, 10–18. Bloomington: Indiana University Press.

Foucault, Michel. 1970. *The Order of Things*. New York: Pantheon.

– 1984. "Nietzsche, Genealogy, History." In *The Foucault Reader*, ed. Paul Rabinow, 76–100. New York: Pantheon.

– 1993. *The Archaeology of Knowledge*. New York: Barnes & Noble.

Frazer, James. 1930. *Myths of the Origin of Fire*. London: Macmillan.

– 1984. "The Fall of Man." In *Sacred Narrative*, ed. Alan Dundes, 72–97. Berkeley: University of California Press.

Freud, Sigmund. 1950. *Totem and Taboo*. New York: W.W. Norton & Company.

– 1960. *The Acquisition and Control of Fire*. London: Hogarth Press. (*The Standard Edition of the Complete Psychological Works of Sigmund Freud*, vol. 22)

– 1964. *New Introductory Lectures on Psycho-Analysis and Other Works*. London: The Hogarth Press.

– 1964b. *Civilization and Its Discontents*. London: Hogarth Press. (*The Standard Edition of the Complete Psychological Works of Sigmund Freud*, vol. 21)

– 1967. *Moses and Monotheism*. New York: Vintage Books.

Freye, Northrop. 1957. *Anatomy of Criticism: Four Essays*. Princeton: Princeton University Press.

Girard, René. 1986. *The Scapegoat*. Baltimore: Johns Hopkins University Press.

– 1987. *Things Hidden Since the Foundation of the World*. Standford: Stanford University Press.

Gleick, James. 1987. *Chaos: Making a New Science*. New York: Viking.

Goswami, Amit, with Maggie Goswami. 1983. *The Cosmic Dancers: Exploring the Physics of Science Fiction*. New York: Harper & Row.

Gould, Stephen Jay. 1989. *Wonderful Life*. New York: W.W. Norton.

– 1991. "The Creation Myths of Cooperstown." In *Bully for Brontosaurus*, 42–59. New York: W.W. Norton.

– 1997. *Full House*. New York: Three Rivers Press.

– 1999. *Rocks of Ages*. New York: Library of Contemporary Thought.

– 2002. *The Structure of Evolutionary Theory*. Cambridge: Harvard University Press.

– 2003. *I Have Landed*. New York: Three Rivers Press.

– 2003b. "The Narthex of San Marco." In *I Have Landed*, 271–84. New York: Three Rivers Press.

Greene, Brian. 2004. *The Fabric of the Cosmos*. New York: Alfred A. Knopf.

– 2011. *The Hidden Reality*. New York: Alfred A. Knopf.

Grey, George. 1906. *Polynesian Mythology*. London: George Routledge & Sons.

Grice, H.P. 1975. "Logic and Conversation." In *The Logic of Grammar*, eds. Donald Davidson and Gilbert Harman, 64–75. Encino: Dickenson Publishing Company, Inc.

Gross, Alan. 1990. *The Rhetoric of Science*. Cambridge: Harvard University Press.

Gross, Paul. 1994. *Higher Superstition: The Academic Left and its Quarrels With Science*. Baltimore: Johns Hopkins University Press.

– 1998. "Bashful Eggs, Macho Sperm, and Tonypandy." In *A House Built on Sand*, ed. Noretta Koertge, 59–70. New York: Oxford University Press.

Gussow, Mel. 1996. *Conversations with Stoppard*. New York: Grove Press.

Guthrie, Stewart Elliott. 1993. *Faces in the Clouds*. Oxford University Press.

Habermas, Jurgen. 1992. *Postmetaphysical Thinking*. Cambridge: MIT Press.

Haldane, J.B.S. 1932. *Possible Worlds*. London: Chatto and Windus.

Haraway, Donna. 1989. *Primate Visions: Gender, Race and Nature in the World of Modern Science*. London: Routledge, Chapman and Hall.

– 1991. *Simians, Cyborgs, and Women: The Reinvention of Nature*. New York: Routledge.

Harris, Marvin. 1974. *Cows, Pigs, Wars & Witches*. New York: Random House.

Hawking, Stephen. 1988. *A Brief History of Time*. Toronto: Bantam Books.

– 1993. *Black Holes and Baby Universes*. New York: Bantam.

Hawking, Stephen, and Leonard Mlodinow. 2010. *The Grand Design*. New York: Bantam Books.

Herrick, James. 2008. *Scientific Mythologies: How Science and Science Fiction Forge New Religious Beliefs*. Downers Grove: IVP Academic.

Hertz, Robert. 1973. "The Pre-eminence of the Right Hand: A Study in Religious Polarity." In *Right & Left*, ed. Rodney Needham, 3–31. Chicago: University of Chicago Press.

Hesiod. 1953. *Theogony*. Indianapolis: Bobbs-Merrill.

Highfield, Roger. 2002. *The Science of Harry Potter*. New York: Viking.

Hofstadter, Douglas. 2007. *I Am a Strange Loop*. New York: Basic Books.

Hölldobler, Bert, and E.O. Wilson. 2009. *The Super-organism: The Beauty, Elegance, and Strangeness of Insect Societies*. New York: W.W. Norton.

Holt, Jim. 1999. "Infinitesimally Yours." *The New York Review of Books*. May 20, 1999, 63–7.

Horgan, John. 1997. *The End of Science*. New York: Broadway Books.

Hume, David. 2007. *An Enquiry Concerning Human Understanding*. Oxford: Oxford University Press.

Hymes, Dell. 1981. "Reading Clackamas Texts." In *Traditional Literatures of the American Indian*, ed. Karl Kroeber, 117–59. Lincoln: University of Nebraska Press.

Hymes, Dell. 1981b. *In Vain I Tried to Tell You*. Philadelphia: University of Pennsylvania Press.

Jastrow, Robert. 1981. *The Enchanted Loom: Mind in the Universe*. New York: Simon & Schuster.

Johansen, J. Prytz. 1954. *The Maori and His Religion*. Copenhagen: Ejnar Munksgaard.

Johnson, Mark. 1987. *The Body in the Mind*. Chicago: University of Chicago Press.

Joseph, Lawrence. 1991. *Gaia*. New York: St Martin's Press.

Jurdant, Baudouin. 1993. "Popularization of Science as the Autobiography of Science." *Public Understanding of Science* 2:365–73.

Kaku, Michio. 2008. *Physics of the Impossible*. New York: Doubleday.

Kant, Immanuel. 1965. *The Critique of Pure Reason*. New York: St Martin's Press. (Based on the 2nd edition, published originally in German in 1787.)

Kapchan, Deborah, and Pauline Turner Strong, eds. 1999. "Theorizing the Hybrid." *Journal of American Folklore* [special issue] 112:445.

Kardiner, Abram. 1945. *The Psychological Frontiers of Reality*. New York: Columbia University Press.

Kennedy, Duncan. "Atoms, Individuals, and Myths." In *Laughing With Medusa*, eds. Vanda Zajko & Miriam Leonard, 233–52. Oxford: Oxford University Press.

Kirk, G.S., J.E. Raven, and M. Schofield. 1983. *The Presocratic Philosophers*. Cambridge: Cambridge University Press.

Koertge, Noretta, ed. 1998. *A House Built on Sand*. New York: Oxford University Press.

Kraft, Chris. 2002. *Flight: My Life In Mission Control*. New York: Plume.

Kranz, Gene. 2001. *Failure Is Not an Option*. New York: Berkley Books.

Krauss, Lawrence. 1996. *The Physics of Star Trek*. New York: HarperCollins.

Kuhn, Thomas. 1985. *The Copernican Revolution*. Cambridge: Harvard University Press.

– 2000. *The Road Since Structure*. Chicago: University of Chicago Press.

Labinger, Jay, and Harry Collins. 2001. *The One Culture? A Conversation about Science*. Chicago: University of Chicago Press.

Lacan, Jacques. 1977. *Ecrits*. New York: W.W. Norton.

– 1991. *The Seminar of Jacques Lacan*, Book 11. New York: Norton.

Lakoff, George. 1987. *Women, Fire and Dangerous Things*. Chicago: University of Chicago Press.

– 1996. *Moral Politics: How Liberals and Conservatives Think*. Chicago: University of Chicago Press.

– 2004. *Don't Think of an Elephant!* White River Junction: Chelsea Green Publishing.

– 2006. *Whose Freedom?* New York: Picador.

– 2006b. *Thinking Points*. New York: Farrar, Straus and Giroux.

– 2008. *The Political Mind: Why You Can't Understand 21st-Century Politics With an 18th-Century Brain*. New York: Viking.

Lakoff, George, and Mark Johnson. 1980. *Metaphors We Live By*. Chicago: University of Chicago Press.

– 1999. *Philosophy in the Flesh*. New York: Basic Books.

Lanier, Jaron. 2010. *You are Not a Gadget*. New York: Alfred A. Knopf.

Latour, Bruno. 1993. *We Have Never Been Modern*. Cambridge: Harvard University Press.

– 2002. "What Is Iconoclash? Or Is There a World Beyond the Image Wars?" In *Iconoclash: Beyond the Image Wars in Science, Religion, and Art*, eds. Bruno Latour and Peter Weibel, 14–37. Karlsruhe: ZKM.

Lederman, Leon. 2002. "Physics, from Leonard to Hertz." In Microsoft *Encarta* Reference Library.

Lee, H.D.P. 1967. *Zeno of Elea*. Amsterdam: Adolf M. Makkert.

Lessl, Thomas. 1985. "Science and the Sacred Cosmos: The Ideological Rhetoric of Carl Sagan." *Quarterly Journal of Speech* 71:175–87.

LeVay, Simon. 2008. *When Science Goes Wrong*. New York: Plume.

Lévi-Strauss, Claude. 1967. "The Structural Study of Myth." In *Structural Anthropology*, 202–28. Garden City: Anchor Books.

– 1967b. "Do Dual Organizations Exist?" In *Structural Anthropology*, 128–60. Garden City: Anchor Books.

– 1969. *The Raw and the Cooked*. New York: Harper & Row.

– 1969b. *Elementary Structures of Kinship*. Boston: Beacon Press.

– 1970. *The Savage Mind*. Chicago: University of Chicago Press.

– 1971. "The Story of Asdiwal." In *The Structural Study of Myth and Totemism*, ed. Edmund Leach, 1–47. London: Tavistock Publications.

– 1976. "How Myths Die." In *Structural Anthropology*, Vol. II. New York: Basic Books.

– 1988. *The Way of the Masks*. Seattle: University of Washington Press.

Levy, Robert. 1990. *Mesocosm*. Berkeley: University of California Press.

Lightman, Alan, ed. 2005. *The Best Amercian Science Writing 2005*. New York: HarperCollins.

Lincoln, Bruce. 1996. "Mythic Narrative and Cultural Diversity in American Society." In *Myth and Method*, ed. Laurie Patton and Wendy Doniger, 163–76. Charlottesville: University Press of Virginia.

– 1999. *Theorizing Myth*. Chicago: University of Chicago Press.

Lindbergh, Anne Morrow. 1969. *Earth Shine*. New York: Hartcourt, Brace & World.

Lindbergh, Charles. 1974. "Foreword" to Michael Collins, *Carrying the Fire: An Astronaut's Journeys*, ix–xiii. New York: Farrar, Straus and Giroux.

Lindsay, Hamish. 2001. *Tracking Apollo to the Moon*. London: Springer-Verlag.

Livy. 1985. *The Early History of Rome*. Harmondsworth: Penguin. Translated by Aubrey de Selincourt.

Lovejoy, Arthur. 1960. *The Great Chain of Being*. New York: Harper & Row.

Lovelock, James. 1995. *The Ages of Gaia*. Oxford: Oxford University Press.

Lowry, Malcolm. 1965. *Under the Volcano*. Philadelphia: Lippincott.

Lucretius. 1951. *On the Nature of the Universe*. London: Penguin. Translated by R.E. Latham.

Lycan, William. 1987. *Consciousness*. Cambridge: MIT Press.

MacLeish, Archibald. 1968. "Riders on the Earth." *The New York Times*. December 25.

Mailer, Norman. 1970. *Of a Fire On the Moon*. Boston: Little, Brown.

Maine, Henry. 1972. *Ancient Law*. London: Everyman's Library.

Malin, David, and Katherine Roucoux. 2002. *Heaven and Earth*. London: Phaidon Press.

Malinowski, Bronislaw. 1944. *A Scientific Theory of Culture*. New York: Oxford University Press.

Margolis, Howard. 2002. *It Started with Copernicus*. New York: McGraw-Hill.

Margulis, Lynn. 1986. *Microcosmos: Four Billion Years of Microbial Evolution*. New York: Summit Books.

Martin, Emily. 1996. "The Egg and the Sperm: How Science has Constructed a Romance Based on Stereotypical Male-Female Roles." In *Feminism & Science*, eds. Evelyn Fox Keller and Helen E. Longino, 103–17. Oxford: Oxford University Press.

McManus, I.C. 2002. *Right Hand, Left Hand: The Origins of Asymmetry in Brains, Bodies, Atoms, and Cultures*. Cambridge: Harvard University Press.

Michael, John. 2000. *Anxious Intellects: Academic Professionals, Public Intellectuals & Enlightenment Values*. Durham: Duke University Press.

Midgley, Mary. 1992. *Science as Salvation: A Modern Myth and its Meaning*. London: Routledge.

Minsky, Marvin. 1986. *The Society of Mind*. New York: Simon & Schuster.

– 2006. *The Emotion Machine*. New York: Simon & Schuster.

Mlodinow, Leonard. 2008. *The Drunkard's Walk: How Randomness Rules Our Lives*. New York: Pantheon.

Moss, Jean Dietz. 1993. *Novelties in the Heavens*. Chicago: University of Chicago Press.

Müller, F. Max. 1874. "Comparative Mythology." In *Chips From a German Workshop*, Vol. II. New York: Scribner, Armstrong, and Co.

Nabokov, P. 1981. *Indian Running*. Santa Fe: Ancient City Press.

Nagy, Gregory. 2002. "Can Myth Be Saved?" In *Myth: A New Symposium*, eds. Gregory Schrempp and William Hansen, 240–8. Bloomington: Indiana University Press.

Needham, Rodney. 1972. "Introduction" to Emile Durkheim and Marcel Mauss, *Primitive Classification*. Chicago: University of Chicago Press.

Needham, Rodney, ed. 1973. *Right and Left*. Chicago: University of Chicago Press.

Nelkin, Dorothy. 1995. *Selling Science*. New York: W.H. Freeman and Company.

– 2004. "God Talk: Confusion Between Science and Religion." *Science, Technology, & Human Values* 29:139–52.

Nietzsche, Friedrich. 1956. *The Birth of Tragedy and the Genealogy of Morals*. Garden City: Doubleday.

Numazawa, K. 1984. "The Cultural Historical Background of Myths of the Separation of Sky and Earth." In *Sacred Narrative*, ed. Alan Dundes, 182–92. Berkeley: University of California Press.

O'Flaherty, Wendy Doniger. 1988. *Other Peoples' Myths*. New York: Macmillan.

Olrik, Axel. 1965. "Epic Laws of Folk Narrative." In Alan Dundes, *The Study of Folklore*, 129–41. Englewood Cliffs: Prentice-Hall.

Orr, Allen. 2002. "The Descent of Gould." *The New Yorker*, September 30.

Our Mr Sun. 1956. Produced by Frank Capra. Distributed by Malibu Video, Inc. VHS.

Parrado, Nando. 2006. *Miracle In the Andes: 72 Days on the Mountain and My Long Trek Home*. New York: Crown Publishers.

Pasachoff, Jay. Review of Emanuel Kerry, *Divine Wind*. *Phi Beta Kappa: The Key Reporter*. Spring 2007.

Pascal, Blaise. 2008. *Pensées and Other Writings*. Oxford: Oxford University Press.

Pigliucci, Massimo. 2005. "The So-called Gaia Hypothesis." *Skeptical Inquirer*. 29(no.3):21–6.

Piller, Charles. 2002. "The Code of the Cosmos." *Los Angeles Times*, July 9.

Pinker, Steven. 1997. *How the Mind Works*. New York: W.W. Norton & Company.

– 2005. "How We May Have Become What We Are." In *Curious Minds: How a Child Becomes a Scientist*, ed. John Brockman, 81–9. New York: Vintage Books.

Pinker, Steven, ed. 2004. *The Best American Science and Nature Writing 2004*. Boston: Houghton Mifflin Company.

Plato. 1981. *Timaeus*. Harmondsworth: Penguin. Translated by Desmond Lee.

– 1985. *The Republic*. New York: W.W. Norton. Translated by Richard Sterling and William Scott.

Poole, Robert. 2008. *Earthrise*. New Haven. Yale University Press.

Price, Huw. 1996. *Time's Arrow & Archimedes' Point*. New York: Oxford University Press.

Primack, Joel, and Nancy Abrams. 2006. *The View from the Center of the Universe*. New York: Riverhead Books.

Propp, Vladimir. 1968. *The Morphology of the Folktale*. Austin: University of Texas Press.

Rabinow, Paul. 1977. *Reflections on Fieldwork in Morocco*. Berkeley: University of California Press.

Radcliffe-Brown, A.R. 1965. *Structure and Function in Primitive Society*. New York: The Free Press.

Rappaport, Roy. 1968. *Pigs for the Ancestors*. New Haven: Yale University Press.

Rees, Amanda. 2001. "Anthropomorphism, Anthropocentrism, and Anecdote: Primatologists on Primatology." *Science, Technology, & Human Values* 26:227–47.

Rees, Martin. 2001. *Our Cosmic Habitat*. Princeton: Princeton University Press.

Reynolds, David West. 2002. *Apollo: The Epic Journey to the Moon*. San Diego: Tehabi Books.

Rosch, Eleanor. 1978. "Principles of Categorization." In *Cognition and Categorization*, ed. Eleanor Rosch and B.B. Lloyd, 27–48. Hillsdale: Lawrence Erlbaum Associates.

Russell, Bertrand. 1948. *Human Knowledge: Its Scope and Limits*. New York: Simon & Schuster.

Russell, Doug. 1993. "Popularization and the Challenge to Science-Centrism in the 1930s." In *The Literature of Science*, ed. Murdo William McRae, 37–53. Athens: University of Georgia Press.

Sagan, Carl. 1980. *Cosmos*. New York: Ballantine Books.

– 1994. *Pale Blue Dot*. New York: Random House.

Sahlins, Marshall. 1976. *Culture and Practical Reason*. Chicago: University of Chicago Press.

Said, Edward. 1993. *Culture and Imperialism*. New York: Knopf.

Schneider, David. 1968. *American Kinship*. Englewood Cliffs: Prentice-Hall.

– 1984. *A Critique of the Study of Kinship*. Ann Arbor: University of Michigan Press.

Schrempp, Gregory. 1983. "The Re-education of Friedrich Max Müller." *Man* 18:90–110.

– 1989. "Aristotle's Other Self." In *Romantic Motives: Essays on Anthropological Sensibilities*, ed. George Stocking, 10–43. Madison: University of Wisconsin Press.

– 1992. *Magical Arrows*. Madison: University of Wisconsin Press.

– 1996. "Folklore and Science: Inflections of 'Folk' In Cognitive Research." In *Journal of Folklore Research* 3:191–206.

– 1998. "Distributed Power: A Theme in Native American Origin Stories." In *Stars Above Earth Below*, ed. M. Bol, 15–27. Pittsburgh: Carnegie Museum.

– 1998b. "Mathematics and Traditional Cosmology: Notes on Four Encounters." *Cosmos* 14:211–26.

– 1998c. "The Demon-Haunted World: Folklore and Fear of Regression at the End of the Millennium." *Journal of American Folklore* 111:247–56.

– 2011. "Catching Wrangham: On the Mythology and the Science of Fire, Cooking, and Becoming Human." *Journal of Folklore Research* 48:109–32.

– *Ms. in progress* (a). Is Lucretius a Saint? A Brief for Canonization.

– *Ms. in progress* (b). Science Fundamentalism: Problems With Remythologizing the Cosmos.

Schwimmer, Eric. 1978. Lévi-Strauss and Maori Social Structure. *Anthropologica* 20:201–22.

Searle, John. 1994. *The Rediscovery of the Mind.* Cambridge: MIT Press.

Shaffer, Elinor. 1998. *The Third Culture: Literature and Science.* Berlin: Walter de Gruyter.

Shea, William. 2000. "Looking at the Moon as another Earth: Terrestrial Analogies and Seventeenth-Century Telescopes." In *Metaphor and Analogy in the Sciences*, ed. Fernand Hallyn, 83–104. Dordrecht: Kluwer Academic Publishers.

Shermer, Michael. 2001. *The Borderlands of Science.* Oxford: Oxford University Press.

Shneiderman, Ben. 2003. *Leonardo's Laptop.* Cambridge: MIT Press.

Shumway, David. 1992. *Michel Foucault.* Charlottesville: University Press of Virginia.

Singh, Simon. 2004. *Big Bang: The Origin of the Universe.* New York: HarperCollins.

Slade, Joseph, and Judith Yaross Lee. 1990. *Beyond the Two Cultures.* Ames: Iowa State University Press.

Smith, S.P., ed. 1913. *The Lore of the Whare-wananga.* Memoir 3. Wellington: The Polynesian Society.

Smolin, Lee. 1997. *The Life of the Cosmos.* New York: Oxford University Press.

Snow, C.P. 1998. *The Two Cultures.* Cambridge: Cambridge University Press.

Sokal, Alan. 1998. "What the Social Text Affair Does and Does Not Prove." In *A House Built on Sand*, ed. Noretta Koertge, 9–22. New York: Oxford University Press.

Sorensen, Roy. 1992. *Thought Experiments.* New York: Oxford University Press.

Spencer, Herbert. 1897. *The Principles of Psychology*, Vol. II. New York: D. Appleton and Company.

Spiro, Melford. 1978. *Burmese Supernaturalism*. Philadelphia: Institute for the Study of Human Issues.

Stocking, George. 1968. *Race, Culture, and Evolution*. New York: The Free Press.

– 1974. *The Shaping of American Anthropology*. New York: Basic Books.

Stoppard, Tom. 1974. *Jumpers*. New York: Grove Press.

Strinati, Dominic. 1997. *An Introduction to Theories of Popular Culture*. London: Routledge.

Tambiah, Stanley Jeyaraja. 1990. *Magic, Science, Religion, and the Scope of Rationality*. New York: Cambridge University Press.

Taylor, John. 1991. *Linguistic Categorization*. Clarendon: Oxford University Press.

Thompson, Stith. 1953. "The Star Husband Tale." *Studia Septentrionalia* 4:93–163.

Thornton, Agathe. 1987. *Maori Oral Literature*. Dunedin: University of Otago Press.

Torey, Zoltan. 2009. *The Crucible of Consciousness*. Cambridge: MIT Press.

Tuan, Yi-fu. 1996. *Cosmos & Hearth*. Minneapolis: University of Minnesota Press.

Turner, Mark. 1987. *Death is the Mother of Beauty*. Chicago: University of Chicago Press.

Turney, Jon. 2001. "Telling the Facts of Life: Cosmology and the Epic of Evolution." *Science as Culture* 10:225–47.

Tylor, E.B. 1929. *Primitive Culture*, Vol. 1. London: J. Murray.

– 1970. *Primitive Culture*, Vol. 2. Gloucester: Peter Smith.

"Unscientific Readers of Science." 1998. *The Economist*, May 9.

Uther, Hans-Jorg. 2004. The *Types of International Folktales*, Part I. Helsinki: Suomalainen Tiedeakatemia, Academia Scientiarum Fennica. (FF Communications No. 284)

Vanhoozer, Kevin. 2010. *Remythologizing Theology*. Cambridge: Cambridge University Press.

Vernant, Jean-Pierre. 1982. *The Origins of Greek Thought*. Ithaca: Cornell University Press.

Veyne, Paul. 1988. *Did the Greeks Believe Their Myths?* Chicago: University of Chicago Press.

Vogler, Christopher. 1992. *The Writer's Journey: Mythic Structures for Screenwriters and Storytellers*. Studio City: Michael Wiese Productions.

Walsh, Lynda. 2006. *Sins Against Science*. Albany: State University of New York Press.

Watson, Lyall. 1984. *Heaven's Breath*. New York: William Morrow and Company, Inc.

Weinberg, Steven. 1984. *The First Three Minutes*. Toronto: Bantam Books.

– 2001. *Facing Up: Science and Its Cultural Adversaries*. Cambridge: Harvard University Press.

Wertheim, Margaret [writer and host]. 1998. *Faith & Reason* [videorecording]. New River Media in association with Five Continents Music Inc. VHS.

White, Michael, and John Gribbin. 1992. *Stephen Hawking: A Life in Science*. London: Viking.

Wilson, E.O. 1998. *Consilience*. New York: Alfred A. Knopf.

Wittgenstein, Ludwig. 1968. *Philosophical Investigations*. Oxford: Basil Blackwell.

Wolfram, Stephen. 2002. *A New Kind of Science*. Champaign: Wolfram Media.

Wrangham, Richard. 2009. *Catching Fire: How Cooking Made Us Human*. New York: Basic Books.

Wright, Robert. 1999. "The Accidental Creationist." *The New Yorker*, Dec. 13.

Index

Best, Elsdon, 115, 117–18, 256n7
Bird-David, Nurit, 267n28
Blumenberg, Hans, 202, 213–15, 270n9, 272n22
bodily experience: and metaphor, 122–37; and personal effects, 188
Boslough, John, 225
Boss, Alan, 242n10
Brenkus, John, 253–4n20
Briggs, Charles, 235n11
Brisson, Luc, 33, 236–7n8, 265n19, 276n12
Brockman, John, 13–15, 235–6n1, 236n4
Brooks, Courtney, 271n16
Brownlee, Donald, 242n10
Butler, Judith, 149
butterfly effect, 37–9, 55, 105

Cabala, 179–80
Capra, Frank, 155–60, 185, 254–5n21, 263n4
cartesian dualism, 31, 79, 162, 167, 184
categories: in Aristotle, 121; category of cause, 130–1; category of class, 127–30; category of time, 131–2; in Kant, 121; origin of, 123–37; and philosophy, 120–30; in Rosch, 146; in Spencer, 122; in Wittgenstein, 145–7
centaurs, 9–11
centric knowledge, 21–34. See also anthropocentrism, ethnocentrism
Cernan, Eugene, 271n17
Chaikin, Andrew, 204–5
chain of being, 23, 66–7, 247n37, 247–8n38

chaos, 37–8; 220; and chaos theory, 37–9, 78–9, 106
city as cosmos, xi–xii, 9
cognitive science, 113, 147, 169–70, 178, 186; and cognitive linguistics, 113; and neural structures, 113, 134–5, 153, 168–70, 184, 258–9n15
Cohen, H. Floris, 245n31
Cohen, I. Bernard, 270n9
Cole, K.C., 177
Collins, Francis, 235n10
Collins, Harry, 235–6n1
Comte, Auguste, 181–4, 265–6n20, 269n5; and schema of progress, 180–4
constants of nature, 36–43, 55, 67–70, 197
cooking and culture, 35, 47–8, 57–8, 61–71; and meat, 47–8, 57–9. See also fire
Copernicus, 4, 22, 27–8, 77–8, 109–12, 120, 150, 169, 175–6, 191–216, 230, 255–6n2, 268n1, 268n4, 269n7, 272n22
Cosgrove, Denis, 270–1n15, 271n19, 272n20
Covey, Stephen, 272–3n24
culture, definition of, 52–4; and leisure, 60; and taste, 42, 57–65; and technology, 48–54
Curnow, Jenifer, 256n6
curve, statistical, 72–105; and baseball batting averages, 75, 82, 96; and driven/passive contrast, 75–92, 102–5; and drunkard's walk, 75–6, 102; and left wall, 74, 76
Cushing, Frank Hamilton, 269n6